Early
NATURAL DISASTERS
Encyclopedias

VOLCANOES

by Samantha S. Bell

Early Encyclopedias

An Imprint of Abdo Reference
abdobooks.com

abdobooks.com

Published by Abdo Reference, a division of ABDO, PO Box 398166, Minneapolis, Minnesota 55439.

Printed in China.
102024
012025

Editor: Christa Kelly
Series Designers: Candice Keimig, Joshua Olson
Production Designer: Ryan Gale

Library of Congress Control Number: 2024938359

Publisher's Cataloging-in-Publication Data

Names: Bell, Samantha S., author.
Title: Volcanoes / by Samantha S. Bell
Description: Minneapolis, Minnesota: Abdo Reference, 2025 | Series: Early natural disasters encyclopedias | Includes online resources and index.
Identifiers: ISBN 9781098296063 (lib. bdg.) | ISBN 9798384917069 (eBook)
Subjects: LCSH: Volcanoes--Juvenile literature. | Natural disasters--Juvenile literature. | Weather--Juvenile literature. | Meteorology--Juvenile literature. | Earth science--Juvenile literature. | Encyclopedias and dictionaries--Juvenile literature.
Classification: DDC 551.21--dc23

CONTENTS

Volcanoes

Volcanoes are openings in Earth's surface. These openings are usually in mountains or hills. They let out molten rock from deep underground. The openings also let out gases.

Eruptions

Molten rock and gases cause pressure to build inside volcanoes. Sometimes the pressure becomes too much. The molten rock and gas come to Earth's surface. They spew out of the volcano. This is called an eruption.

During eruptions, some volcanoes spew up to 26,000 gallons (98,000 L) of molten rock each second.

Active, Dormant, and Extinct

Volcanoes that have erupted recently are called active volcanoes. But not all volcanoes are active. Some volcanoes are dormant. These volcanoes have erupted before. Scientists believe they will erupt again. Other volcanoes are extinct. They have not erupted in a long time. Scientists believe they will not erupt again.

There are about 170 active volcanoes in the United States.

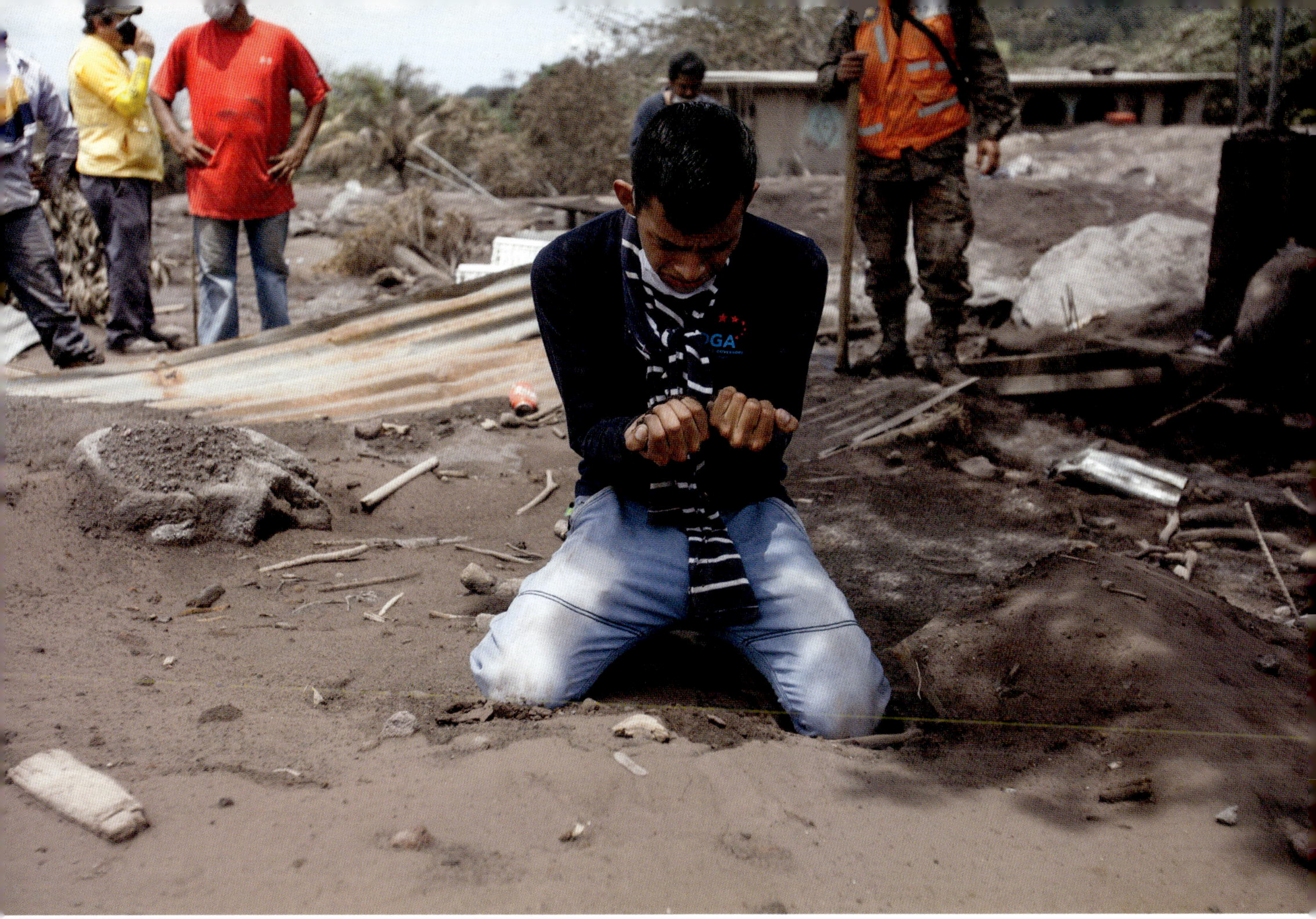

Volcanoes kill more than 500 people every year.

Dangerous Volcanoes

Eruptions can be dangerous. Poisonous gases can make it hard for people to breathe. Molten rock can destroy homes. Volcanic eruptions can destroy property and kill people.

There are more than 1,300 active volcanoes on Earth.

Earth's Layers

Earth is made of layers. The center is called the core. The middle layer is called the mantle. The outer layer is called the crust.

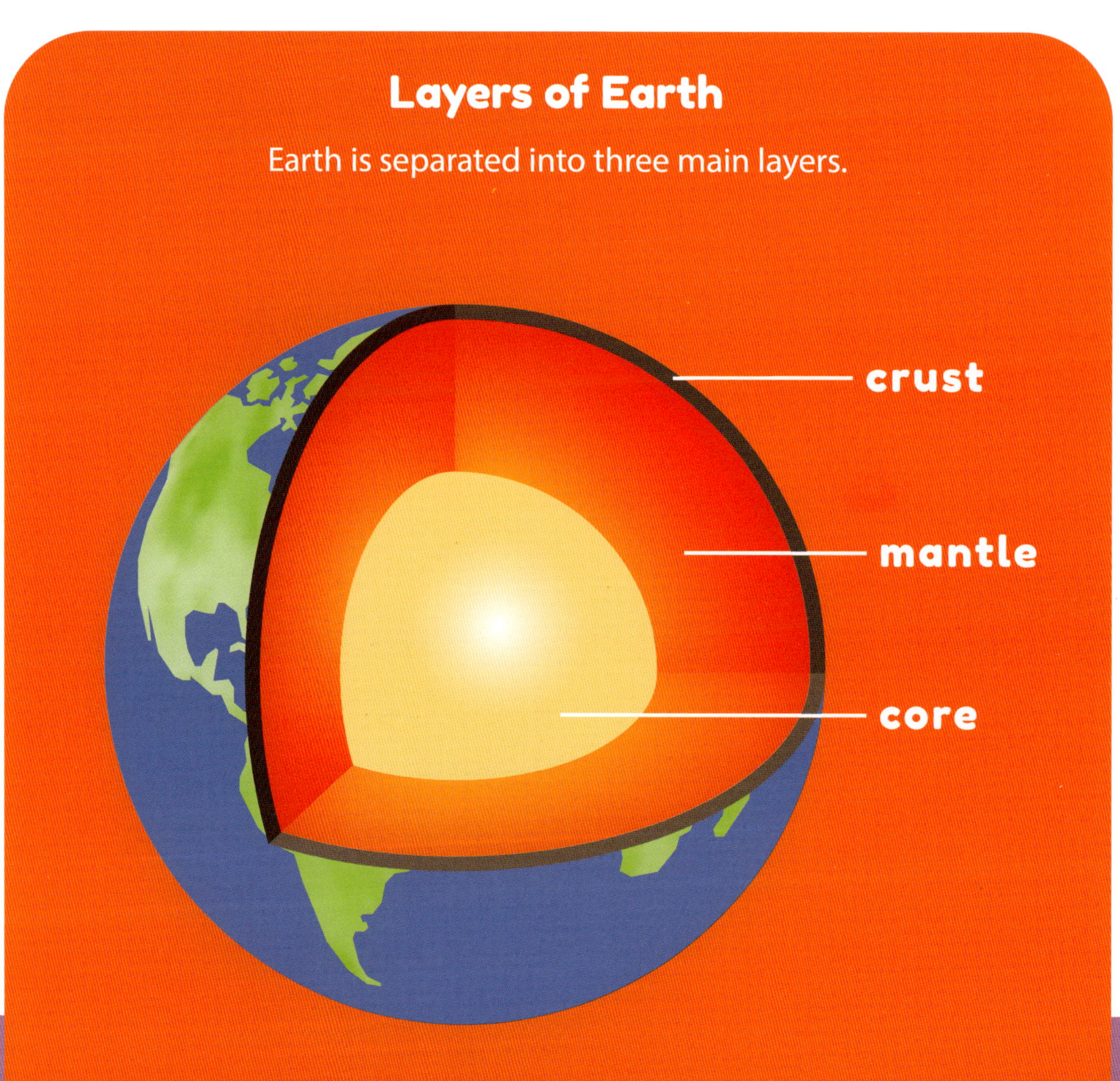

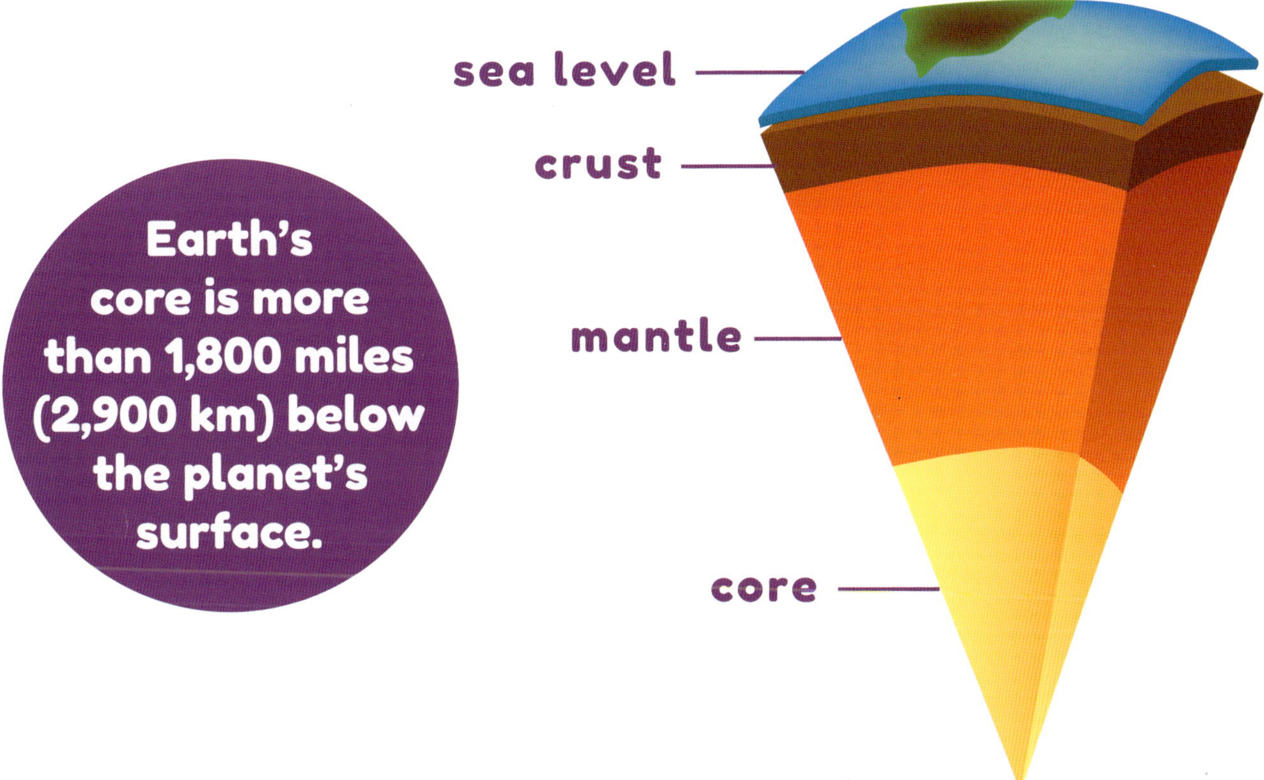

Hot in the Center

Earth's core is very hot. Some parts are up to 10,800 degrees Fahrenheit (6,000°C). The mantle is very hot too. Some parts are more than 6,600 degrees Fahrenheit (3,600°C). Most of the mantle is solid. But the mantle's hot temperature combined with pressure changes can cause rocks to melt. The melted rock becomes a thick, flowing substance. This substance is called magma.

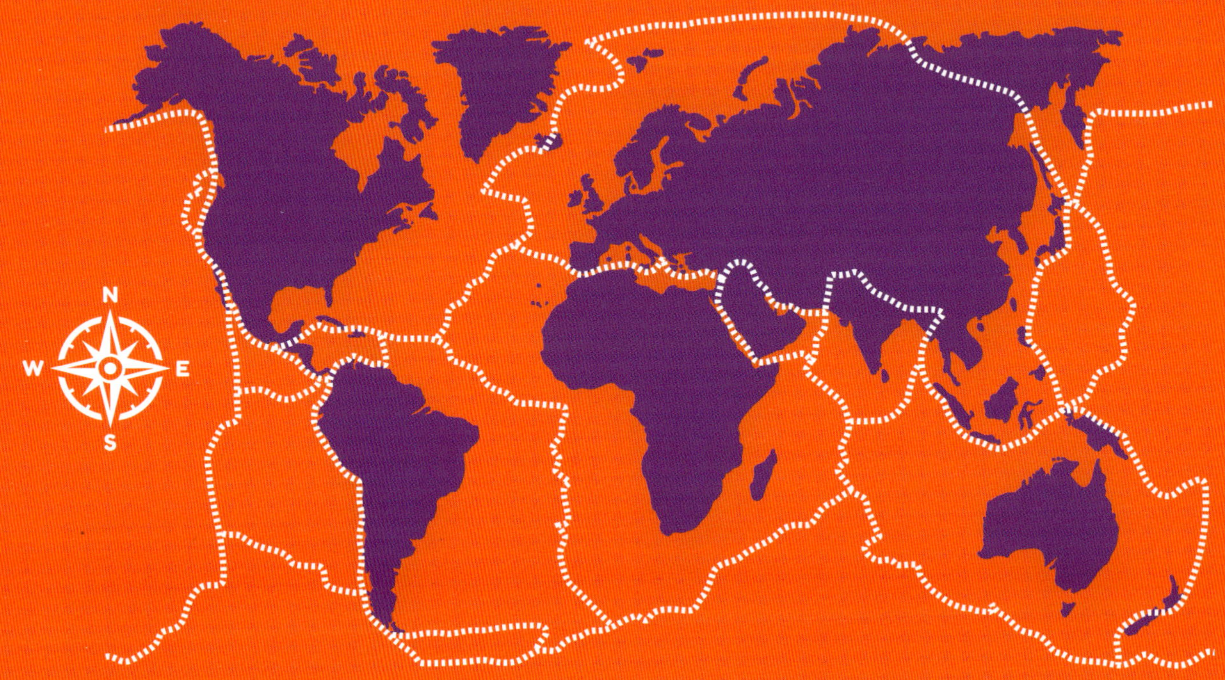

Tectonic Plates

Earth's crust is made of huge pieces called tectonic plates. These plates can span whole continents. They fit tightly together. Tectonic plates slowly move. As they move, they create volcanoes.

Moving Apart

Tectonic plates leave a gap as they move apart. Hot magma rises into that gap. It cools. It makes new crust. But the new crust is thin. Magma can explode through it. This can create a volcano.

Tectonic plates that move away from each other have a divergent plate boundary.

Tectonic plates that push against each other have a convergent plate boundary.

Crashing Plates

Sometimes tectonic plates push against each other. One plate may sink under another. This is called subduction. The sinking plate is pushed far underground. Earth's mantle warms the plate. The rock melts. This creates magma. It also releases gases. The magma and gas travel to the surface. This can create a volcano.

Hotspots

A few volcanoes form in the middle of tectonic plates. This happens when there is magma close to Earth's surface. The magma pushes through the crust to make a volcano. These areas are called hotspots.

Submarine Volcanoes

Most volcanoes are found on the ocean floor. These are called submarine volcanoes. Most of them are thousands of feet underwater.

Hawaii's volcanoes were formed by a hotspot.

How Volcanoes Grow

Volcanoes begin as openings in Earth's surface. But some grow into huge mountains. This growth is caused by eruptions. Eruptions begin in Earth's mantle. Heat and pressure melt some of the rock in the mantle. The rock turns into magma.

It can take hundreds of thousands of years for volcanoes to grow into mountains.

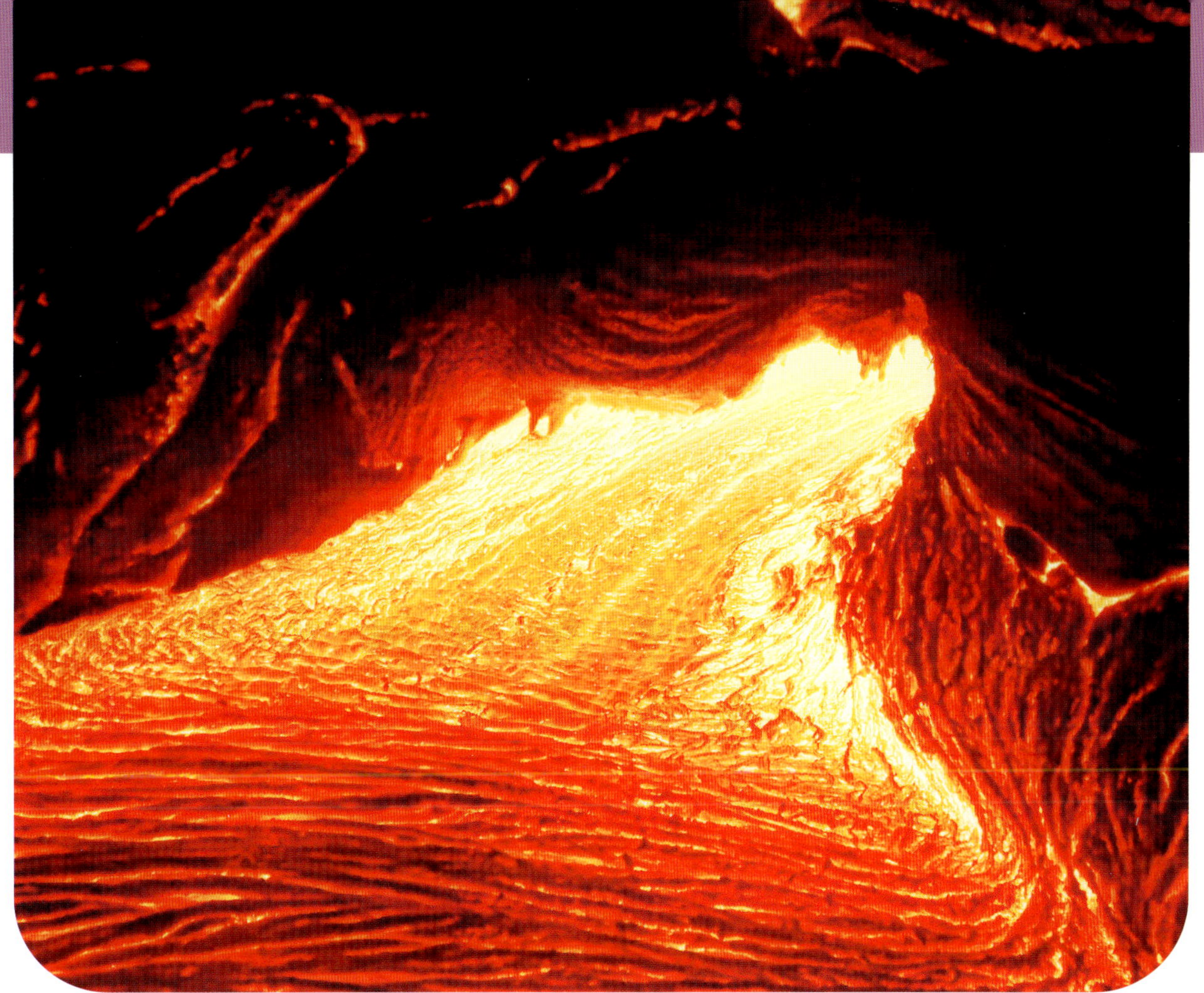

Pressure and high temperatures keep magma in a liquid state.

Gases

Magma contains gases. These gases include water vapor. They also include carbon dioxide and sulfur. The gases make the magma lighter. It weighs less than the solid rock around it. This causes the magma to move upward.

Explorers found a dormant magma chamber in Iceland. It covers an area of more than 35,000 square feet (3,300 sq m).

Into the Chambers

Magma flows up into openings in Earth's crust. These areas are called magma chambers. The chambers are still far below Earth's surface.

Building Up

Magma builds up in the chambers. It pushes upward. It creates a path to Earth's surface. This path is called a conduit. Some people also call it a main vent.

Inside a Volcano

Volcanoes have a magma chamber, a conduit, and at least one vent.

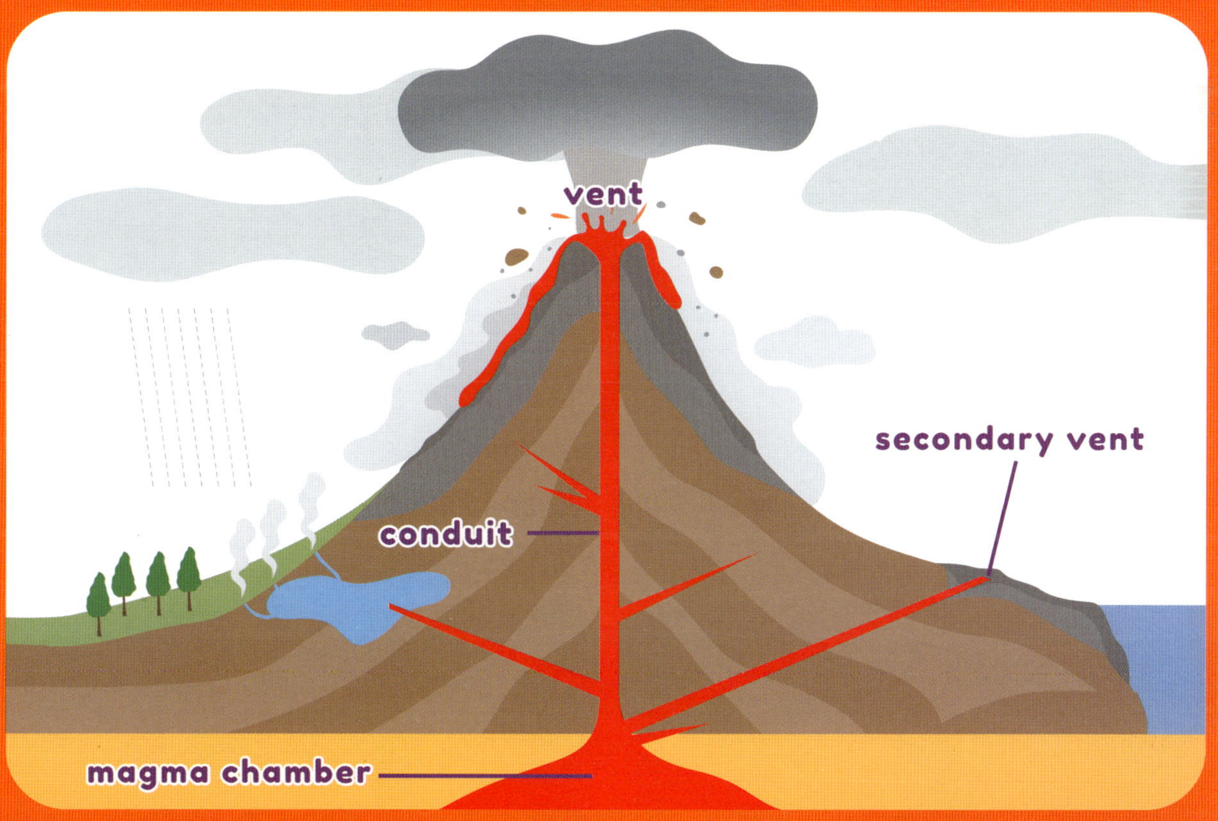

Breaking Through

Gas and magma gather in the conduit. Pressure builds. It becomes too much. The magma finally breaks through the surface. The volcano erupts. The magma and gas spill out a hole at the top of the conduit. This hole is called a vent.

Every day, about 20 volcanoes erupt.

After an eruption, the conduit often fills up with hardened magma, creating a structure called a volcanic plug.

Secondary Vents

Magma and gas may also exit from a smaller conduit. This conduit branches off from the main conduit. It is called a secondary vent.

Some eruptions are so explosive that they blow the tops off volcanoes.

Explosive Volcanoes

Sometimes volcanoes erupt explosively. Other times eruptions are gentle. The amount of silica in the magma determines the explosiveness of an eruption.

Silica

Silica is found in magma. It is one of the main materials in sand. It makes magma thick and slow moving. This makes it harder for gas to escape the magma. When gas cannot escape, pressure builds. Finally, the pressure releases in an explosion. Magma with little silica is runny and fast. The gas can escape into the air bit by bit. The volcano will erupt more gently.

Volcanoes near convergent plate boundaries usually have magma with a lot of silica.

Gas

Erupting volcanoes can release several different materials. One of the most common materials released is gas. This gas can be dangerous. The chemicals in the gas can make it hard to breathe. Areas with gas may also be hazy. This can make it hard to see.

Volcanic gas can be dangerous to plants, animals, and people.

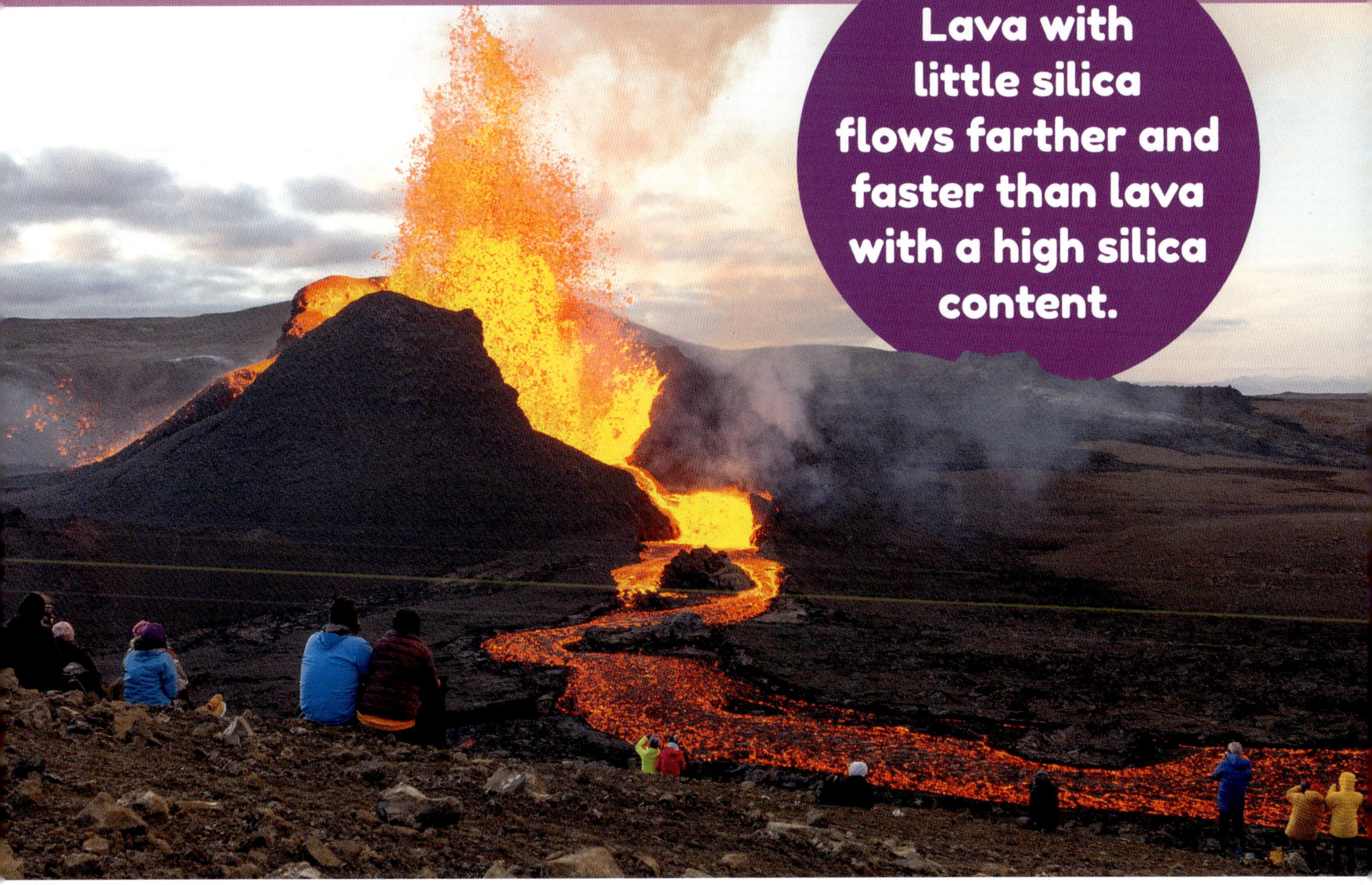

Lava with little silica flows farther and faster than lava with a high silica content.

Lava

Magma that comes to Earth's surface is called lava. Volcanoes may release lava during eruptions. It spills out of a volcano's vent. This is called a lava flow.

Lava can reach 2,282 degrees Fahrenheit (1,250°C).

Pieces of tephra can fall as far as 50 miles (80 km) away from a volcano.

Tephra

In explosive eruptions, magma shoots into the air. It solidifies and breaks into pieces. The pieces are called tephra. Tephra pieces can be as big as houses.

Blocks and Bombs

Tephra pieces bigger than 2.5 inches (6.4 cm) are called blocks and bombs.

Volcanic Ash

Tephra can blast into tiny pieces. Pieces smaller than 0.08 inches (2 mm) are called volcanic ash. The ash can cover huge areas. This is called ashfall.

Volcanic ash pieces can be smaller than 0.0001 inches (0.003 mm).

Lahars

Sometimes tephra mixes with water. This happens when tephra flows into a stream or mixes with melted snow or ice. The tephra turns into mud. The mud moves quickly down the sides of the volcano. This is called a lahar. Some lahars destroy entire communities.

In Indonesia, Mount Agung's lahars have killed hundreds of people.

Pyroclastic flow can move at speeds of 100 miles per hour (160 kmh).

Pyroclastic Flow

During the most dangerous eruptions, material released from the volcano combines into a dangerous mass. Lava flows from the volcano, carrying tephra and rocks. Volcanic gas and ash form hot clouds above the lava. This is called a pyroclastic flow. It destroys anything in its path.

Mount Stromboli has been erupting almost nonstop for more than 2,000 years.

Pulses and Phases

Volcanoes rarely erupt all at once. Eruptions usually have several explosions of magma and gas. Each explosion is called an eruptive pulse. Each pulse can last for minutes at a time. A series of eruptive pulses is known as an eruptive phase.

Long Eruptions

Eruptions can last for days or weeks. Some last for years or centuries. Their eruptive phases can be up to three months apart. But each phase may still be considered part of a single eruption.

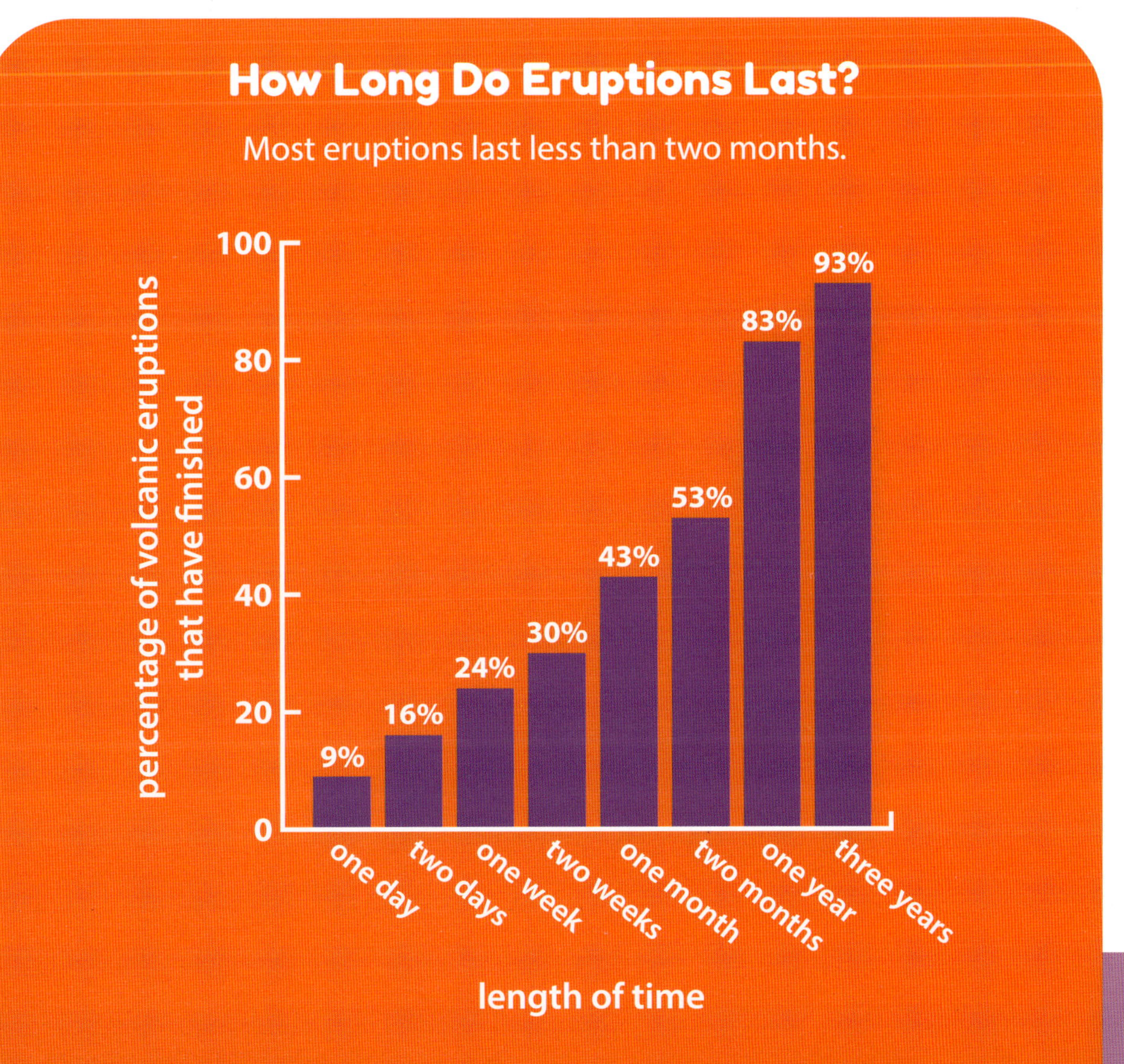

Cooling Off

Lava is very hot when it erupts. It cools over time. An average flow takes about eight months to 1.5 years to cool. The thickest flows may not cool for up to 20 years.

FUN FACT!

Cold temperatures, rain, and wind can make lava cool faster.

When lava cools, it can create massive tubes called pyroducts.

Lava and tephra can build volcanoes into huge mountains.

Building Volcanoes

Lava becomes hard as it cools. It turns into rock. As more eruptions happen, more rock is left behind. Tephra is left behind too. This creates the shape of a volcano.

Each volcano erupts in its own way.

Types of Volcanoes

Volcanoes can take different shapes. The way they erupt determines their shapes. There are three common types of volcanoes.

Cinder Cones

Cinder cones are the most common type of volcano. They are usually short and steep. They have a bowl-shaped area at the top. This is called the crater. Cinder cones form when tephra explodes from a volcano. It falls around the vent. This forms the volcano.

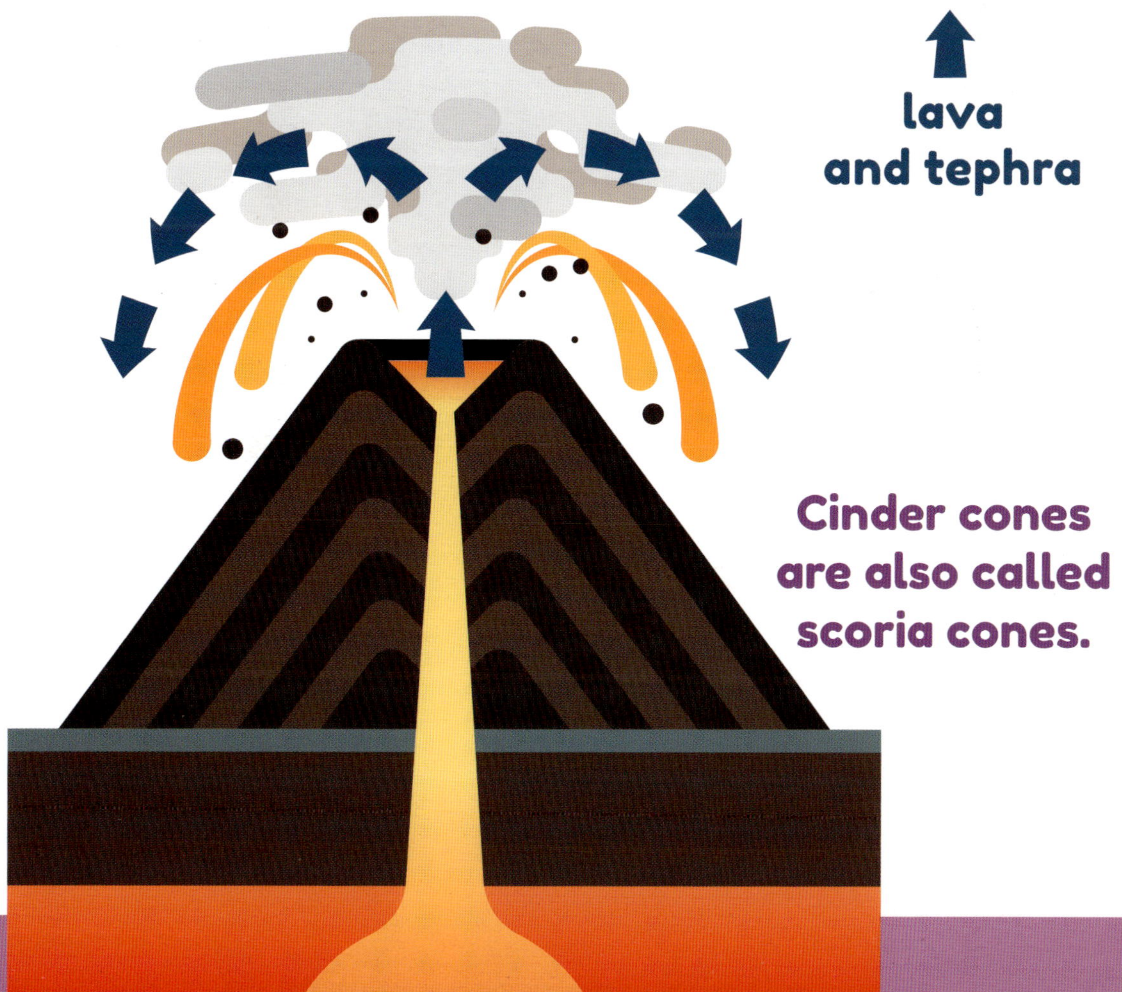

Cinder cones are also called scoria cones.

Composite Volcanoes

Composite volcanoes are usually very tall. They are formed in layers. The layers are made of slow-moving lava flows, ash, and rock. More layers are added each time the volcano erupts.

Some people call composite volcanoes stratovolcanoes.

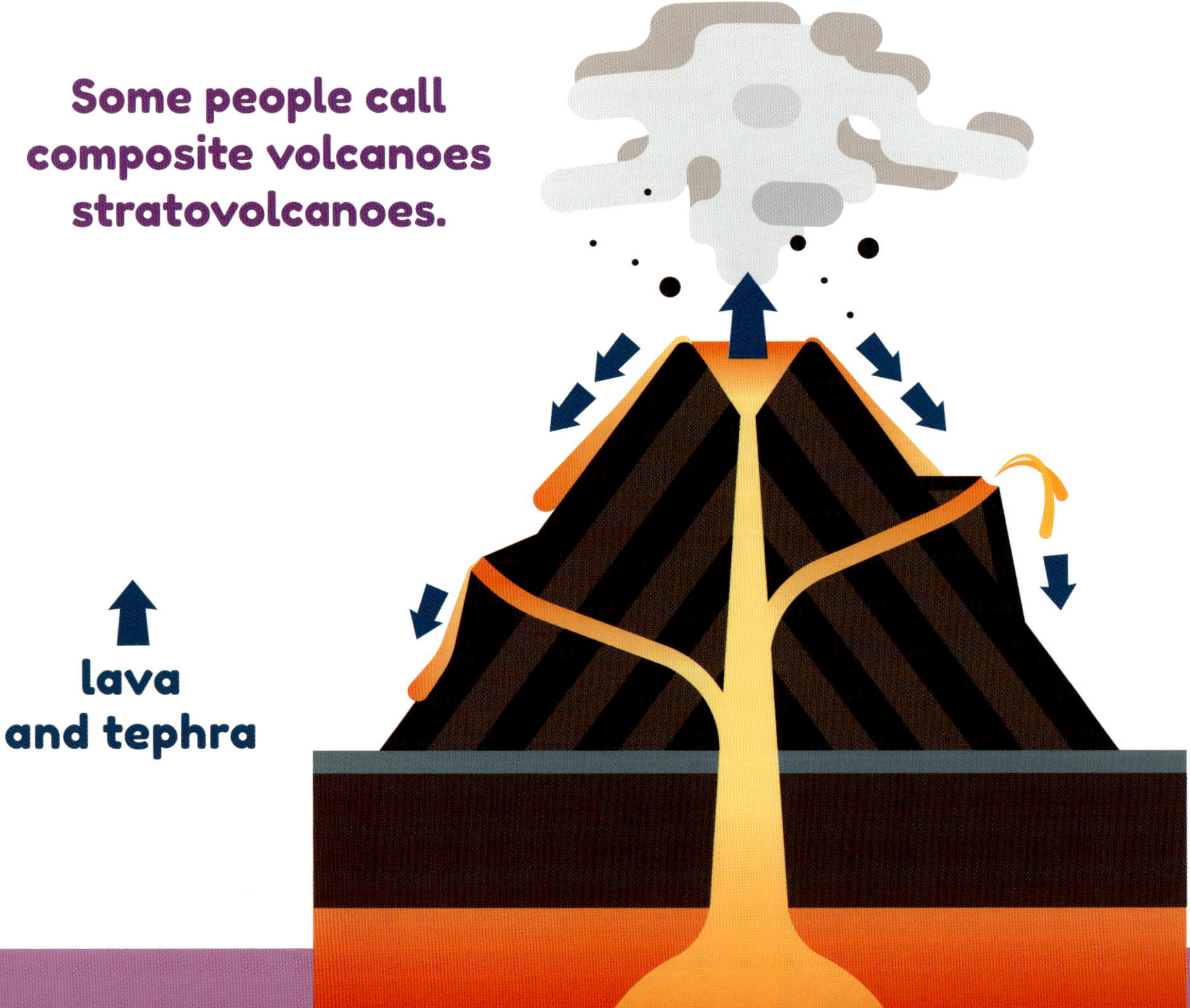

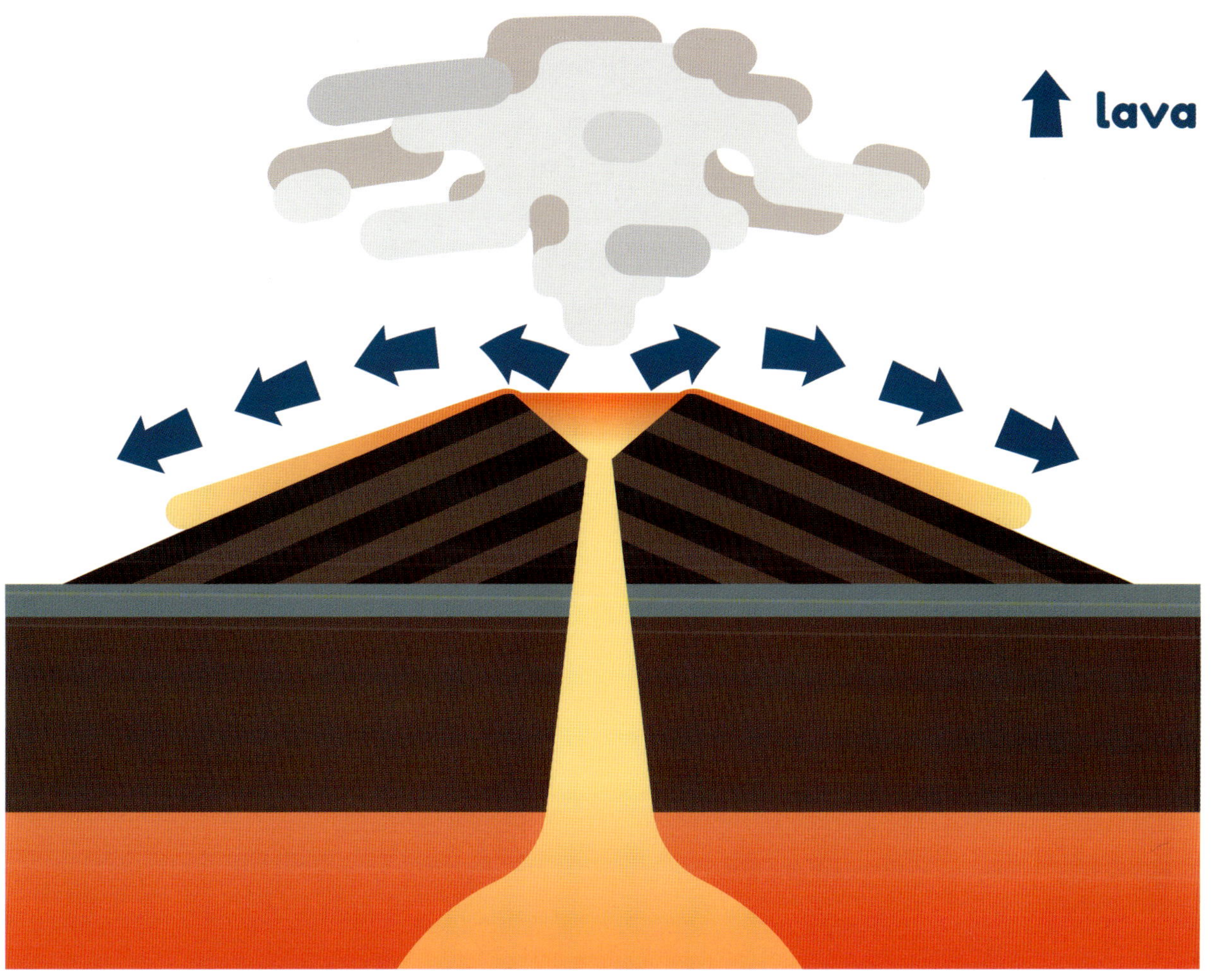

Shield volcanoes can stay active for more than one million years.

Shield Volcanoes

Shield volcanoes build up slowly. They are formed by fast-moving lava. The lava spreads out across a large area and cools. It forms a slightly sloping hill.

In 2015, 59 million people lived within six miles (10 km) of an active volcano.

Why Predict Eruptions?

Volcanic eruptions can destroy whole cities. They can kill many people. Scientists are learning how to predict eruptions. This lets them warn people ahead of time. This can save many lives.

Volcanologists

Scientists who study volcanoes are called volcanologists. They try to predict volcanic eruptions. Some travel to study volcanoes up close. Others work at observatories.

The United States has five volcano observatories. The first was founded in 1912.

People who want to be volcanologists can get degrees in geology, chemistry, or another science-related field.

Difficult Job

It is not easy to predict eruptions. Scientists cannot see what is happening underground. They have to study what is on the surface. And there is much they do not know about volcanoes.

Dangerous Work

Working near a volcano can be dangerous. Scientists use special safety equipment. They wear gas masks. Their gear must be able to resist heat.

Volcanologists travel around the world to study volcanoes.

Though long-term forecasting can help make predictions, it is often unreliable. Volcanoes do not always follow patterns.

Long-Term Forecasting

One type of prediction is long-term forecasting. This is when scientists try to predict eruptions months or years in the future. Scientists study a volcano's history. They research past eruptions. They determine how often it erupts. This can give them clues about when it may erupt again.

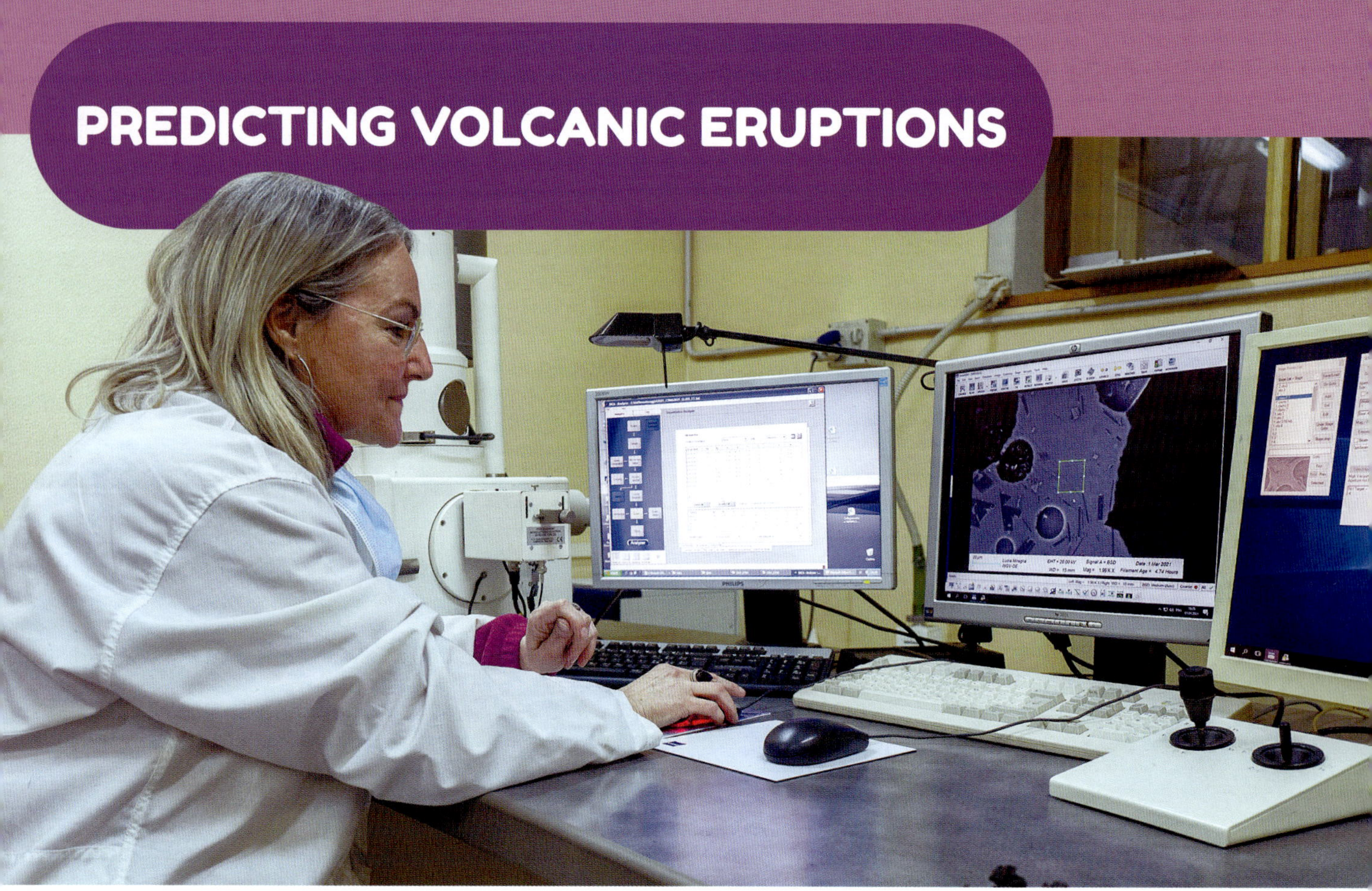

Scientists can study volcanic rock and ash samples to determine when a volcano last erupted.

Studying Past Eruptions

Scientists may also study what a volcano's past eruptions were like. They look at the rock left behind by past eruptions. This can tell them about the type of eruption a volcano might have. This helps them predict how dangerous future eruptions may be.

Mapping

Scientists also study the area around a volcano. They mark the locations of cooled lava. This is called mapping. Mapping helps scientists find the areas that were affected by past eruptions. This helps them predict the areas that may be affected in the future.

Satellites can map eruptions as they are happening.

Making Maps

People use many different methods to map volcanoes. They study older maps. They examine photos taken from planes. They look at satellite images. They visit volcanoes. They combine this information to make accurate maps.

Maps of volcanoes can show which areas may be in danger during an eruption.

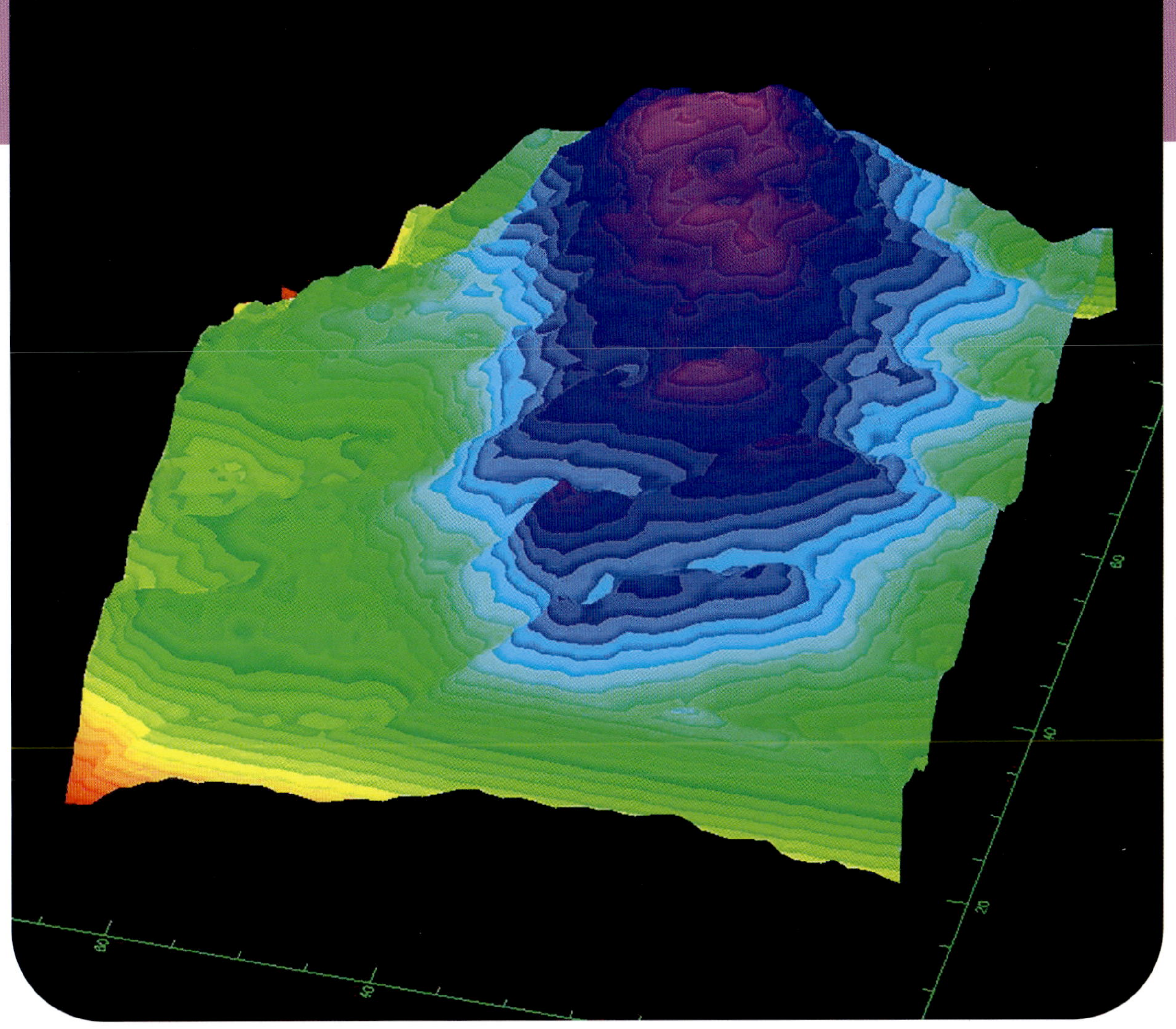

Radar maps help scientists prepare for eruptions.

Radar Mapping

Scientists can also use radar mapping instruments to map volcanoes. Aircraft and satellites carry these tools. They create 3D maps of Earth's surface. They help scientists predict where lava flows or lahars might travel.

Short-Term Forecasting

Scientists also do short-term forecasting. This helps them predict eruptions that will occur soon. They may be able to warn people weeks ahead of time. Other times, they may be able to warn people only a few hours before an eruption.

Scientists are getting better at forecasting eruptions.

Successful eruption forecasts can save thousands of lives.

Changes in Activity

Volcanologists watch for changes in volcanoes. They compare any changes with past eruptions. They see if the activity is similar. This helps them predict eruptions.

In 2017, hundreds of earthquakes were detected around Mount Agung before its eruption.

Earthquakes

Earthquakes are one change scientists look for. When magma moves underground, it can cause earthquakes. It can also make new cracks in the ground. These changes may mean an eruption is coming.

Seismographs

Scientists use instruments called seismographs to detect earthquakes. These machines measure how strong an earthquake is. They can also tell where an earthquake is. The machines monitor earthquakes every day. They send the information they gather to scientists' computers.

History of the Seismograph

The first seismograph was created in Italy in 1875.

The taller the line made by the seismograph, the stronger the earthquake.

Ground Swelling

Volcanologists also look for ground swelling. Before an eruption, magma collects under the ground. This makes the surface of the ground swell. Satellites can often detect this swelling.

In 2022, the ground around Mauna Loa swelled before the volcano erupted.

Strainmeters are often installed near volcanoes and places with frequent earthquakes.

Tiltmeters and Strainmeters

Scientists can use instruments to detect ground swelling. One instrument scientists use is a tiltmeter. It measures the tilt or slant of the ground. It can detect the change in tilt when the ground swells. Strainmeters can help as well. They measure changes in the shape of Earth's crust.

FUN FACT!

Strainmeters can even detect ground changes caused by the weather.

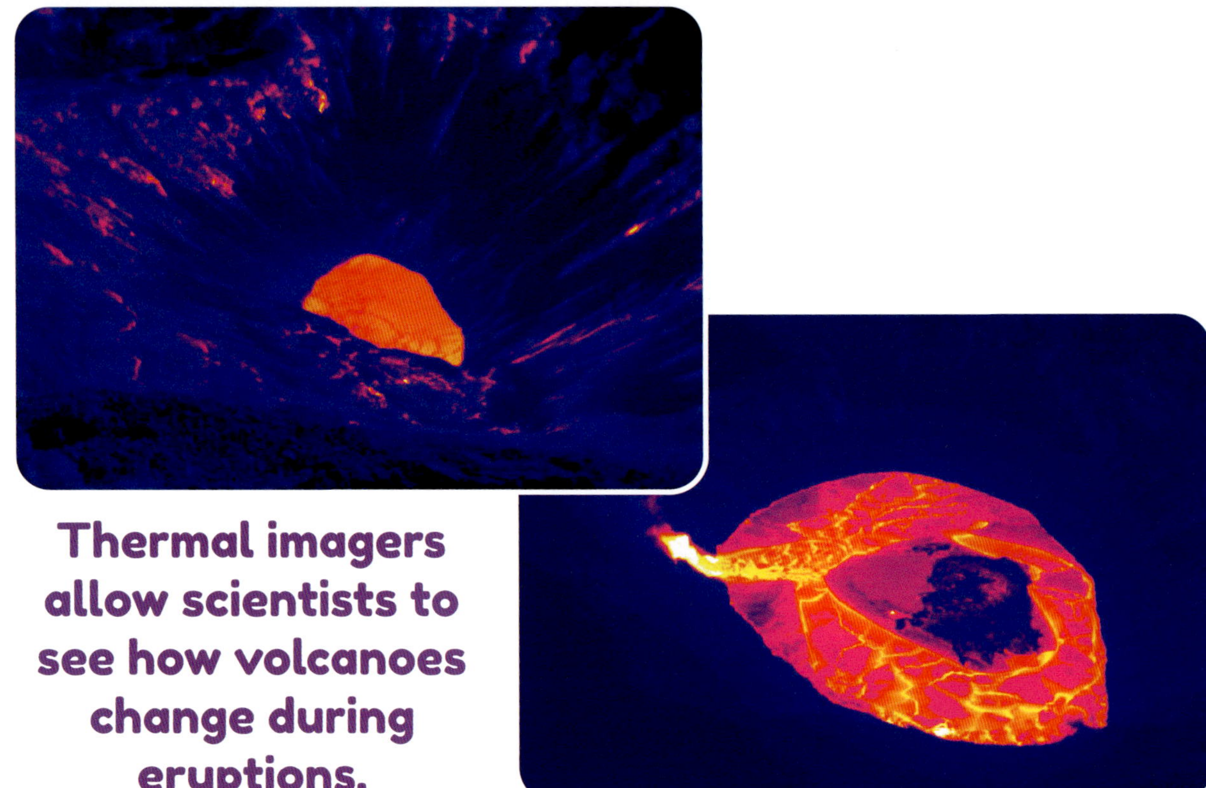

Thermal imagers allow scientists to see how volcanoes change during eruptions.

Thermal Imagers

Scientists can also use thermal imagers to predict eruptions. Thermal imagers are special cameras. They detect heat. Then they create an image. The images show the temperature of different objects. Warmer areas often appear yellow or orange. Cooler areas appear blue or purple.

Images from Above

Aircraft and satellites carry thermal imagers. The imagers take pictures of volcanoes. Scientists study the photos. They can see if the ground temperature changes. This could signal that a volcano is about to erupt.

Taking thermal images from planes and satellites can be safer than taking images from the ground near an active volcano.

Lava Flows

A lava flow can destroy anything in its path. It may knock over buildings and trees. It can light homes on fire and bury large areas.

In 2018, lava flows in Hawaii destroyed more than 700 homes.

Pahoehoe and aa are the two main types of lava flow. Pahoehoe, *left*, is smooth. Aa, *right*, is rocky.

Unusable Land

Lava flows can cover huge areas of land. When the lava cools, the land is buried under rock. People usually cannot use the land anymore. They must move their homes and businesses.

When lava flows into a body of water, it can cause the water to boil.

Lahars can travel at speeds of 50 miles per hour (80 kmh).

Faster than Lava

Lahars can cover huge amounts of land too. They usually travel down river valleys. They pick up speed as they go. They can move faster than lava. They may reach people far away from a volcano.

Deadly Lahars

Lahars can bury buildings and land. They can destroy bridges and roads. This can trap people. The lahars may be too deep or hot for people to cross. Lahars can kill many people.

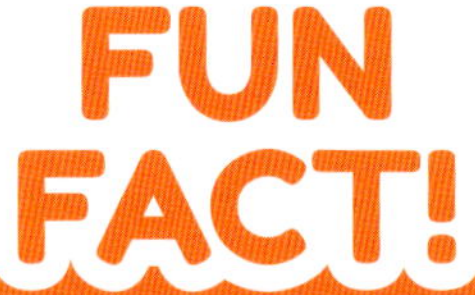

Lahars can be hotter than 200 degrees Fahrenheit (100°C).

Most deaths from volcanic eruptions are caused by lahars and pyroclastic flow.

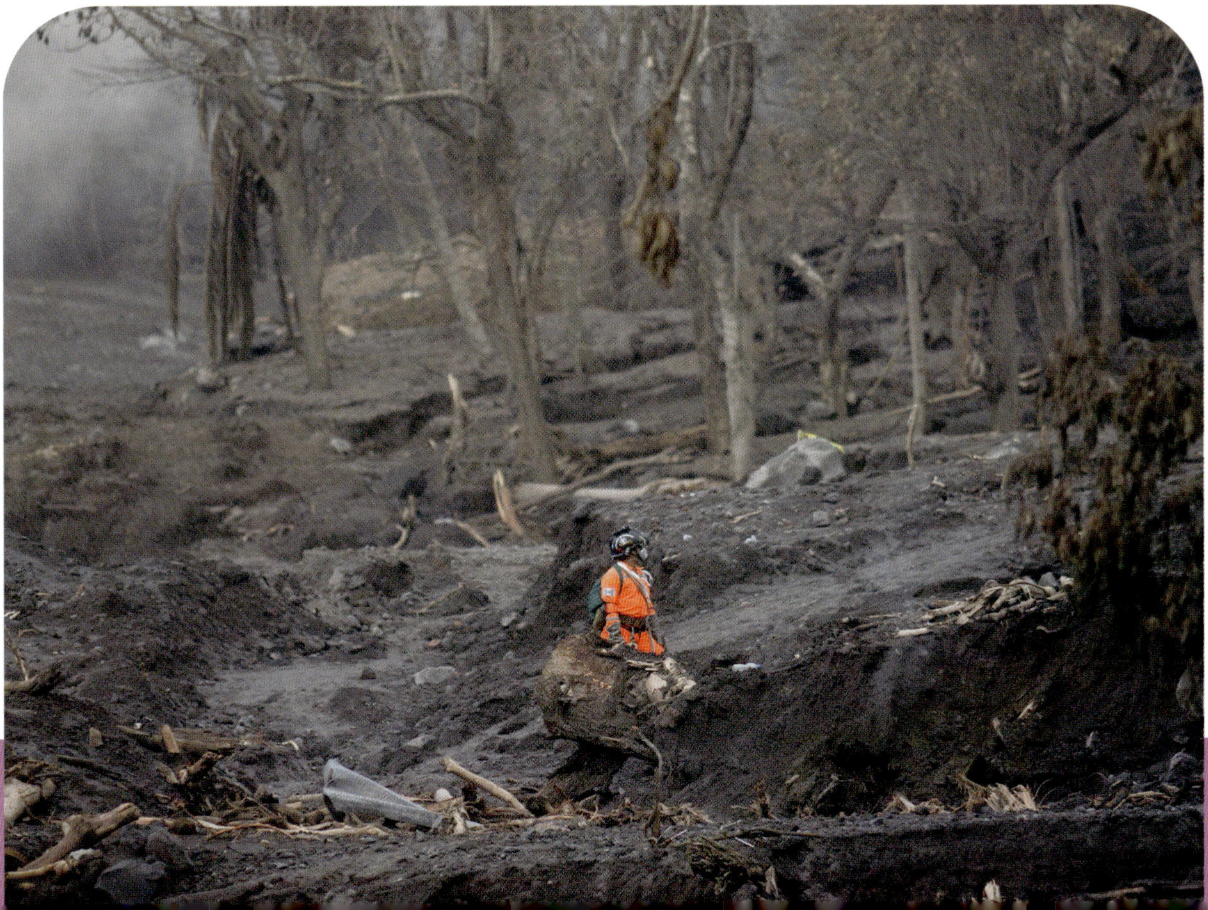

Large eruptions can release more than 20,900 pounds (9,500 kg) of ash every second.

Ashfall

Volcanic ash can be dangerous too. Communities near volcanoes are at the greatest risk. But ash particles can be carried by the wind. Ashfall can affect people hundreds of miles away.

Breathing In Ash

Volcanic ash can cause many health problems. It is especially dangerous for babies and elderly people. It is also dangerous for people with respiratory conditions. The particles may contain crystalline silica. This can cause lung disease. It can also scar the lungs.

Volcanic ash can make it hard to breathe.

Eye Problems

Ash can also cause eye problems. Ash from wood is soft and fluffy. But volcanic ash is made of rocks. It is sharp. The sharp edges can scratch or tear the front of the eye. It can make eyes itchy or bloodshot.

Volcanic ash particles are sharp and dangerous.

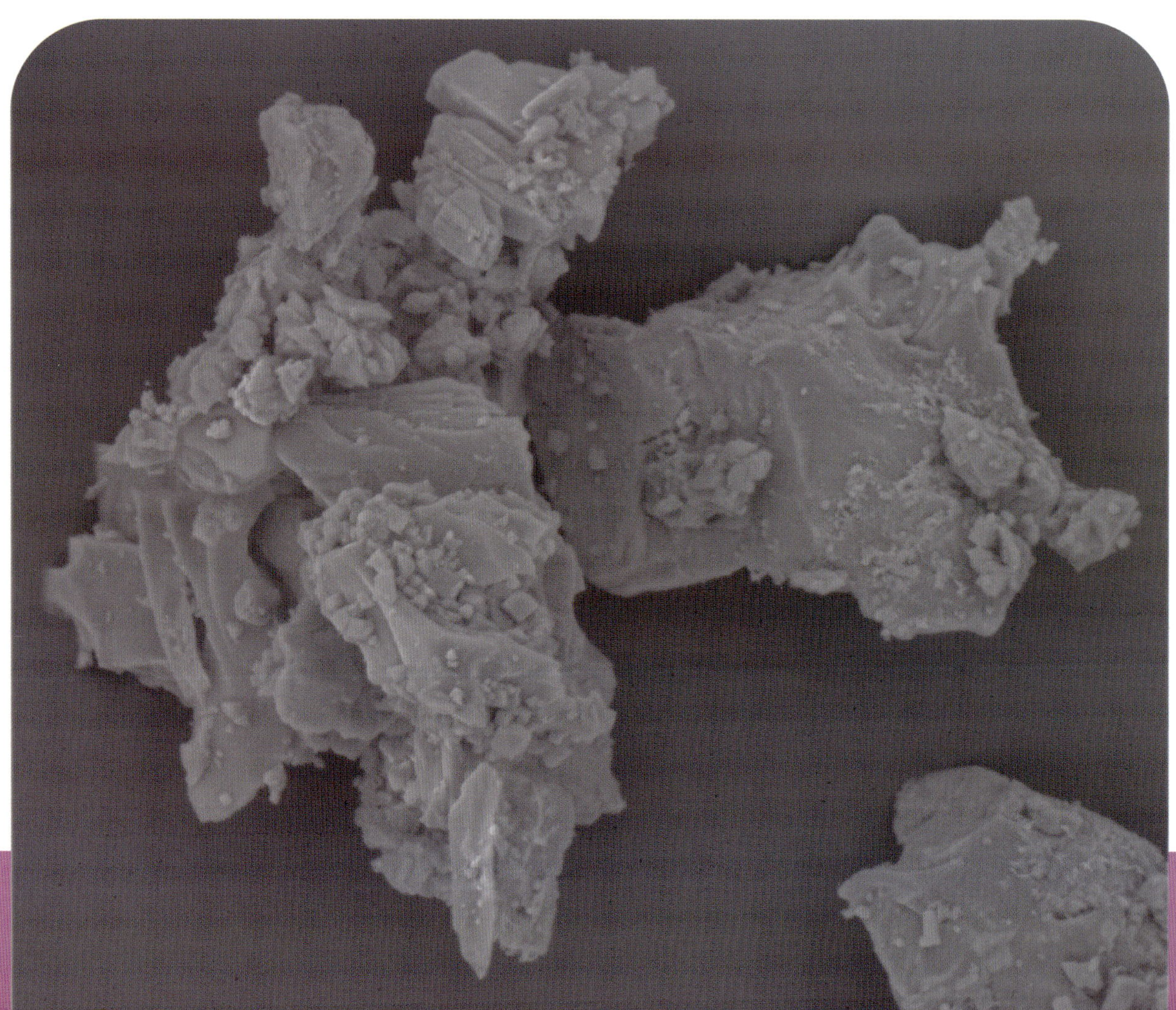

In 1982, volcanic ash shut down all four of a passenger plane's engines. The crew got the engines working again just in time.

Ash and Planes

Volcanic ash is also dangerous for aircraft. The ash covers the sky. It makes everything dark and hard to see. It can also damage planes. Volcanic ash contains particles of rock and glass. These can get inside an airplane's engine. The particles can cause the engine to stop working.

Volcanic gases can make water acidic and dangerous.

Poisonous Gases

Harmful gases can come from volcanoes before, during, and after an eruption. These gases may come in contact with skin or eyes. This can cause mild reactions, such as itchy skin. Sometimes people breathe the gases in. This can make it hard to breathe. Poisonous gases can kill animals and people.

Acid Rain

Rainfall can mix with volcanic gases. This makes acid rain. Acid rain is rain that carries dangerous chemicals. The chemicals can hurt people. They can kill plants. They can damage property.

Hawaii's Kilauea volcano releases gas that causes acid rain.

Landslides

Eruptions can also cause landslides. Hot magma and gases can cause the steep tops of volcanoes to break away. The rock slides down the volcano. This is called a landslide.

Large landslides may turn into lahars.

Thousands of people are killed in landslides every year.

Landslides can bury entire towns.

Changing Landscapes

Landslides move very fast. They often change an area's landscape. Landslides can cover valleys with debris. They can form small hills. They can close off streams and create lakes. They can even bury people.

In 2021, lava from Spain's Cumbre Vieja volcano started fires. The fires destroyed nearly 3,000 buildings.

Natural Disasters

Volcanic eruptions often cause other natural disasters. The rising magma and gases can cause earthquakes. Earthquakes can happen before, during, or after an eruption. Eruptions also often cause wildfires. Lava can start fires. So can hot ash.

Tsunamis

Sometimes eruptions create giant waves called tsunamis. Landslides can cause tsunamis. So can underwater eruptions. Tsunamis can crash into the shore. They devastate coasts and drown people.

In 2022, the underwater Hunga Tonga-Hunga Ha'apai volcano erupted, causing a massive tsunami.

Making New Islands

Volcanic activity does not always destroy land. Sometimes it makes new land. Underwater eruptions can create new islands. Lava erupts from underwater volcanoes. The lava builds up in layers. The layers may eventually reach the surface of the water. A new island is formed.

FUN FACT!

Most volcanic eruptions occur underwater

Volcanoes have been forming Hawaii for more than 70 million years.

Building an Island

Islands may be built up over many eruptions.

Pressure builds in an underwater volcano.

The underwater volcano erupts.

Lava cools and hardens on top of the volcano.

Over time, layers of lava build up. This creates new land.

Built over Time

Volcanic islands can be any size. Some islands form quickly. They may take just hours or days to build. Others may continue to form for years.

A Volcanic State

The US state of Hawaii is made of islands formed by volcanoes. Two volcanoes are still forming the Big Island.

THE EFFECTS OF VOLCANOES

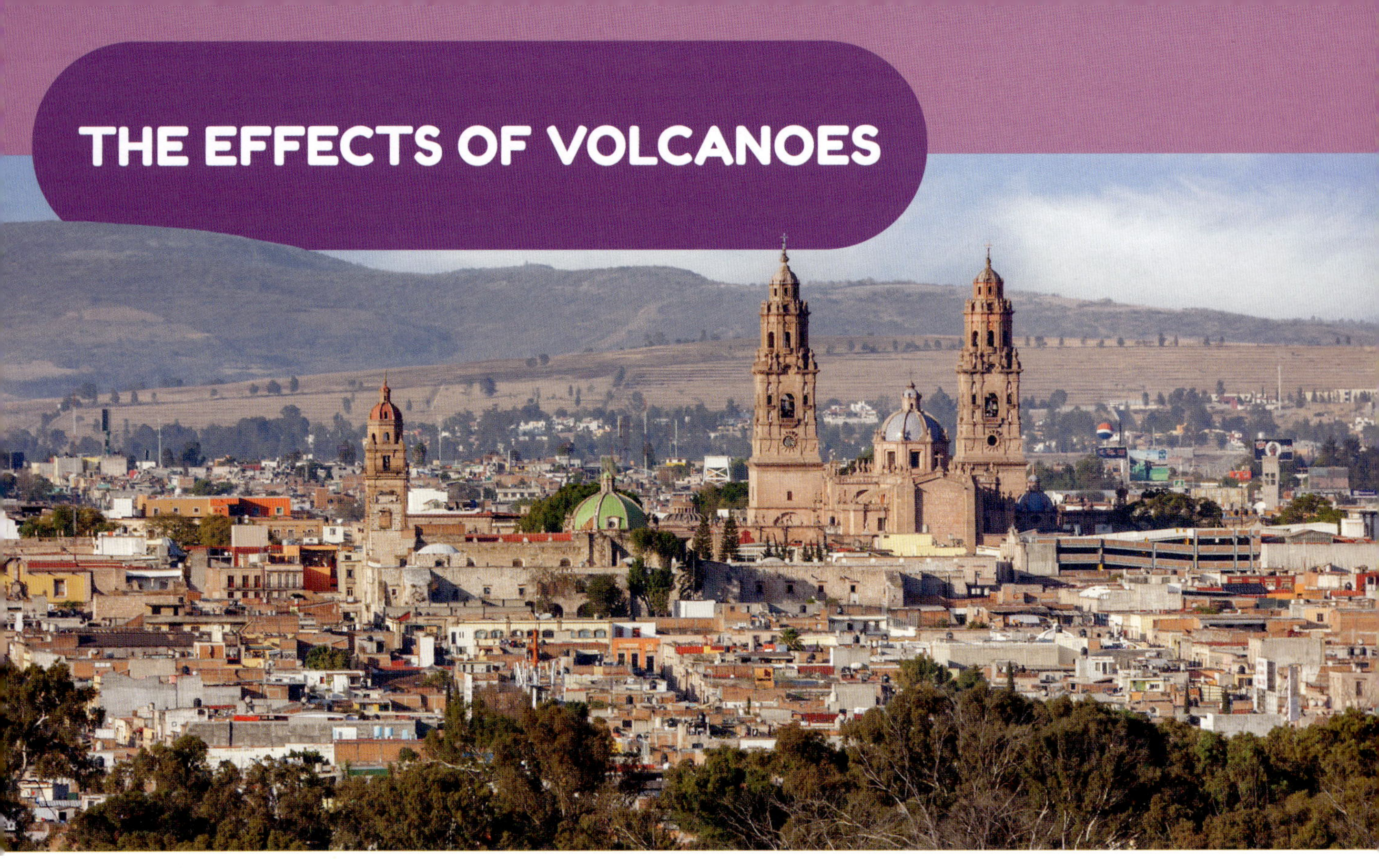

Nearly six million people live within three miles (5 km) of Mexico's Michoacán-Guanajuato volcanoes.

Living near a Volcano

Volcanoes can be very dangerous. But people still choose to live near them. This is because they also provide many benefits.

FUN FACT!

The Cascade mountain range in the United States and Canada was formed mainly by volcanoes.

Volcano Benefits

The ash from eruptions makes soil fertile. Crops grow well. Farmers can produce a lot of food. Volcanoes also create tourism. People like to visit volcanoes. This brings money to local shops and restaurants.

Millions of tourists travel to see volcanoes every year.

Providing Energy

Volcanoes can provide energy. Magma produces heat. The heat causes water in the ground to become hot. The water's heat can be turned into electricity. This is called geothermal energy.

How Does Geothermal Energy Work?

Hot water is pulled out of the ground. Its steam turns turbines and generates electricity. A cooling tower turns the steam back into water. The water is then put back into the ground.

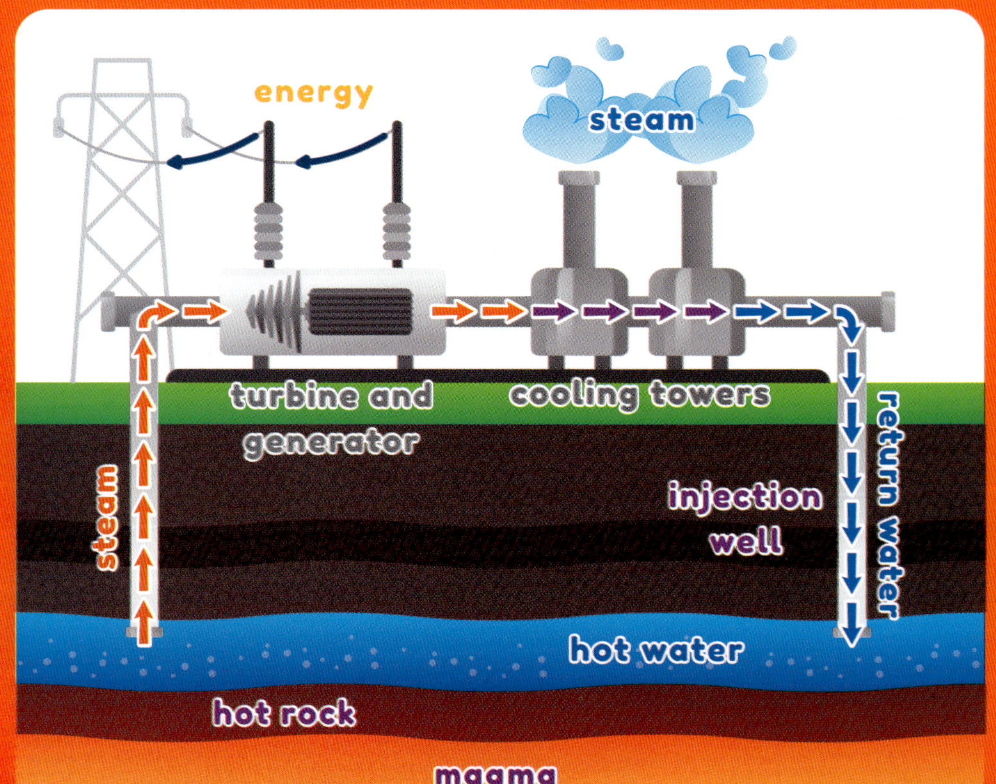

Three million people in the United States use geothermal energy.

Renewable Energy

Geothermal energy is a renewable energy source. That means it never runs out. Heat is always being produced inside Earth.

Emergency Supplies

Even people who do not live near volcanoes should have an emergency supplies kit.

Preparing Supplies

The first step to staying safe during a volcanic eruption is to be prepared. Collecting supplies is an important part of preparation. Emergency supplies should include food, water, and medicine. People should also gather first aid supplies and a flashlight. Batteries and chargers should be available as well.

Evacuation Kit

Supplies can be stored in a kit. People who live near volcanoes should have a kit for evacuating. The evacuation kit should have supplies for at least three days. But it should be small enough that someone can carry it.

Ready to Go

Sometimes people have weeks of warning before they evacuate. But other times, people may have only a few hours. Evacuation kits can help them leave quickly.

In addition to normal emergency supplies, people who live near volcanoes should have masks and goggles to protect them from ash.

Home Kit

People who live near volcanoes should also have a home kit. This kit contains supplies in case a person is stuck at home after an eruption. It may take a while for rescue teams to arrive. The kit should have enough supplies to last two weeks. It should have enough medicine to last one month.

People who live near volcanoes should have an emergency supply of 14 gallons (53 L) of water per person.

Evacuation Routes

It is also important for people near volcanoes to memorize their evacuation routes. Officials determine these routes. Evacuation routes help people avoid dangers, such as areas at risk for lava flows or lahars.

The United States Geological Survey website contains live information about eruptions in the United States.

Watching the News

People who live near volcanoes should stay informed about eruptions. They can watch the local news. These stations provide updates about nearby volcanoes.

Alert Levels

Many countries use alert levels to indicate the current conditions of a volcano. People should pay attention to these alerts. The United States has four alert levels. US volcano observatories determine a volcano's alert level. People can sign up to receive alerts through emails. They can also see alerts on the observatories' websites.

Volcano observatories around the world work together to protect people from eruptions.

Normal and Advisory

The first alert level is *normal*. This level means a volcano is not showing signs of eruption. The next alert level is *advisory*. This means a volcano is showing signs of unrest.

Scientists might declare an advisory if minor earthquakes occur around a volcano.

US Volcanic Alert Levels

The United States has four volcanic alert levels.

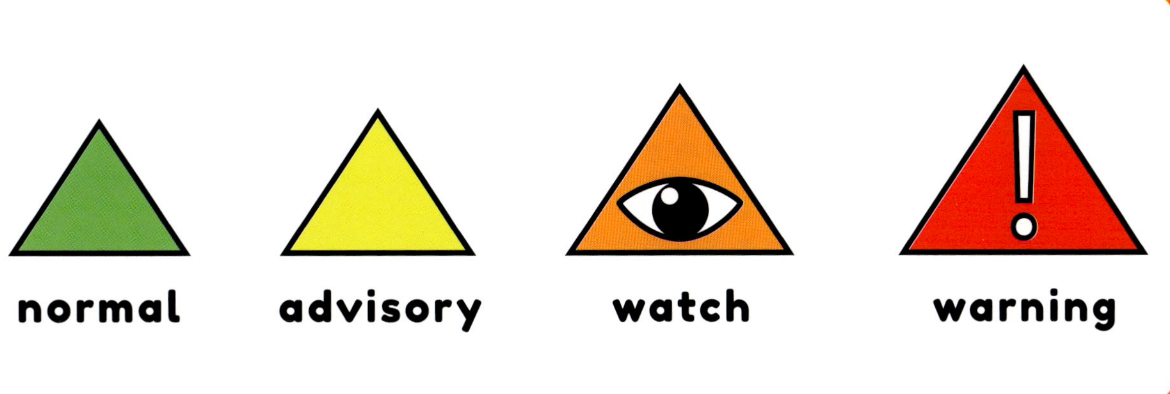

Watch and Warning

Watch is the next alert level. This means signs of volcanic activity are increasing. There may be an eruption in the future, but scientists do not know when it will happen. Watch can also mean that a volcano is erupting but is not posing a big danger. The highest alert level is *warning*. This means that a dangerous eruption will happen very soon. It may also mean an eruption is currently happening.

The Aviation Color Code system was developed in the 1990s by the Alaska Volcano Observatory.

Warning Airplanes

Different countries have different volcanic alert level systems. But they share the same alert level system to warn airplanes about eruptions and ash. Sharing a single system helps airplanes stay safe when they fly to different countries.

Aviation Color Code

The alert level system for airplanes is called the Aviation Color Code. The system has four levels. Each level is represented by a color.

Scientists often consult with each other before changing a volcano's Aviation Color Code.

Color Levels

Green means that a volcano is quiet. Yellow means there is some activity. Orange means an eruption is likely. It may also mean that an eruption is happening but is not creating ash. Red means that an eruption will occur soon. It may also mean that an eruption is happening and is releasing ash.

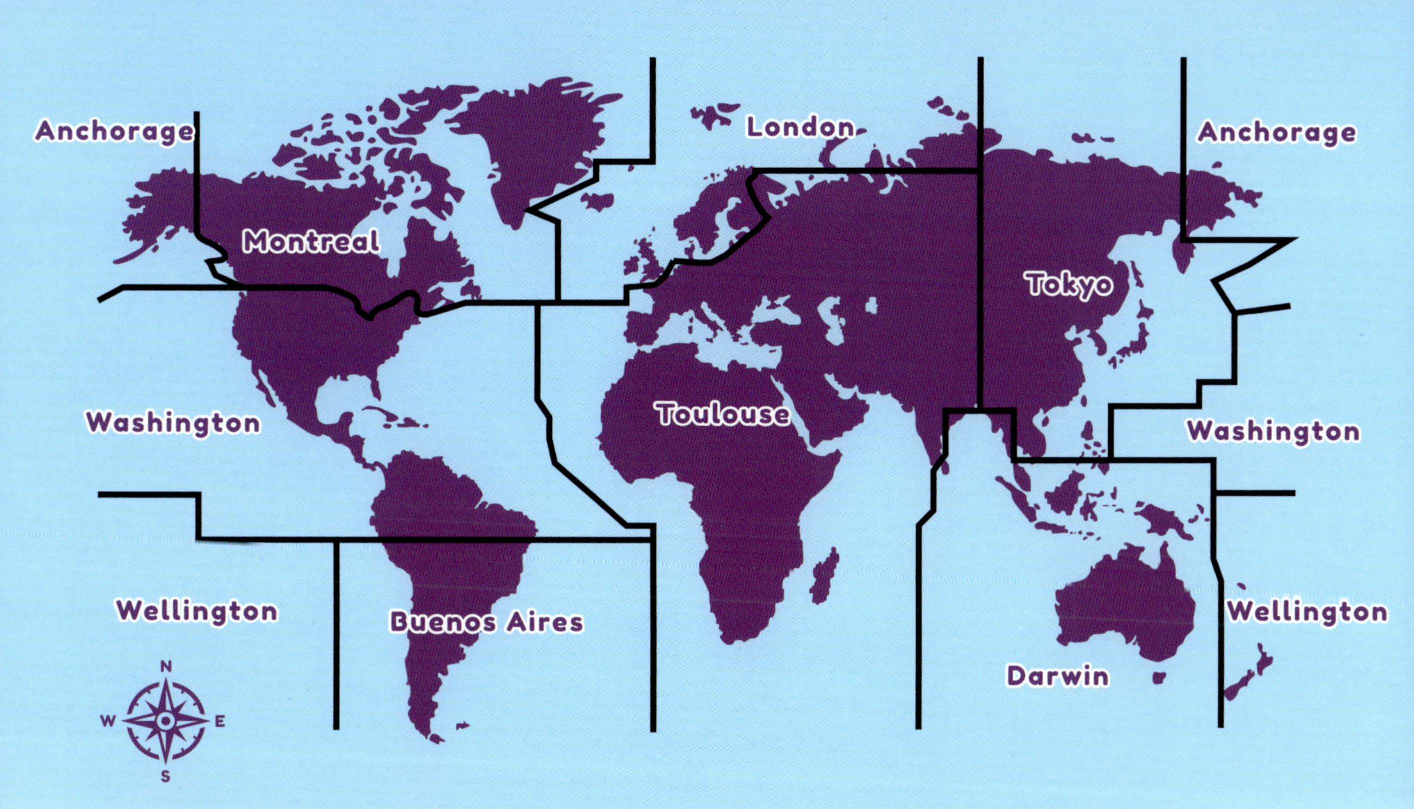

Each Volcanic Ash Advisory Center monitors ash in a specific part of Earth's airspace.

Keeping Track

There are nine Volcanic Ash Advisory Centers around the world that determine aviation color levels. Each observatory monitors ash within its airspace. The scientists in the observatories update the color codes.

In 2023, nearly 18,000 people in the Philippines were evacuated after Mayon Volcano erupted.

Evacuating

When people are alerted that a volcano will erupt, they must follow instructions given by officials. Sometimes officials decide it is too dangerous for people to stay in their homes. They order people to evacuate. This means that everyone must leave the area. People should take their evacuation kits and move to safety.

Sheltering at Home

Other times, officials will tell people to stay home. People must prepare their homes. They should close all doors and windows.

People who are unable to shelter at home should find shelter with a family member or locate a public disaster shelter.

Blocking All Openings

People should cover any openings to their homes. They can place damp towels at the base of doors to keep ash from coming inside. People should also cover any fireplace openings and turn off all heating and air conditioning units.

Placing tape or plastic over window seams can keep out ash.

Taking Care of Animals

People must also take care of their animals. Pets should be brought indoors. Livestock should be kept in closed shelters.

If possible, people should shelter in a room without windows.

Water

During an eruption, people should drink only bottled water. Regular drinking water can become contaminated by volcanic ash.

Stay Inside

Once an eruption begins, people should stay inside. They should not open any doors or windows. This could let ash inside. Officials will tell people when it is safe to go outside again.

Going Outdoors

People should not go outside during an eruption. But if they must go, they should wear a mask. This keeps them from breathing in ash. It is also important to wear protective clothing. This includes long-sleeved shirts and long pants. Goggles can help protect the eyes from ash.

Wearing protective clothing helps prevent ash from hurting the skin, eyes, and lungs.

Areas to Avoid

People should avoid areas downwind of the eruption. These areas receive the most ash. It is also important to stay away from low-lying areas that may receive lahars.

Masks

People should wear N95 respirator masks if they go outside during an eruption. These masks filter out ash particles. N95 masks are available at hardware stores.

Traveling by car is dangerous during eruptions. Lahars can overwhelm cars, and ash can clog engines.

After eruptions, it can take months or years for areas to be safe for return.

After the Eruption

Conditions may still be dangerous after an eruption. People who evacuated should wait to return. Officials will tell them when it is safe to come back. People who stayed home should stay away from areas with lava flow and ash.

Ash is heavy. If it is not cleaned off houses, it might cause roofs to collapse.

Cleaning Up

Cleaning up volcanic ash can take a long time. It can take weeks or months to remove it all. Spraying water on the ash sometimes helps. This keeps the ash from blowing back into the air. But too much water can make ash heavy and sticky. Then it is hard to remove.

Removing Tephra

Once roads are safe to drive on, big trucks will move tephra to the sides of roads. Diggers then pick up tephra and put the pieces into dump trucks. The tephra is then taken to a landfill. This process can take months.

Cleaning up tephra is expensive and time consuming.

Cooling the Climate

The climate is the weather in an area over time. An eruption can have a large effect on the climate. It can cause temperatures to cool.

Eruptions can affect the climate of areas thousands of miles away.

Ash can stay in the air for more than a year after an eruption.

Ash in the Atmosphere

Some eruptions send huge amounts of ash into the atmosphere. Many of these particles fall back to Earth. But the smallest particles can remain in the air.

Sunlight can be used to make electricity.

Warming Earth

The sun sends sunlight and energy to Earth. This is called solar radiation. It warms Earth's surface. The warm surface then heats the atmosphere.

Blocking the Sun

Ash can reflect solar radiation. When ash is in the air, some of the sun's solar radiation bounces off the ash particles. This stops sunlight from reaching the ground. Ash can also absorb radiation. It soaks up some of the heat. Both of these effects cause Earth to cool.

How Ash Cools Earth

Ash can reflect solar radiation. This makes the world cooler.

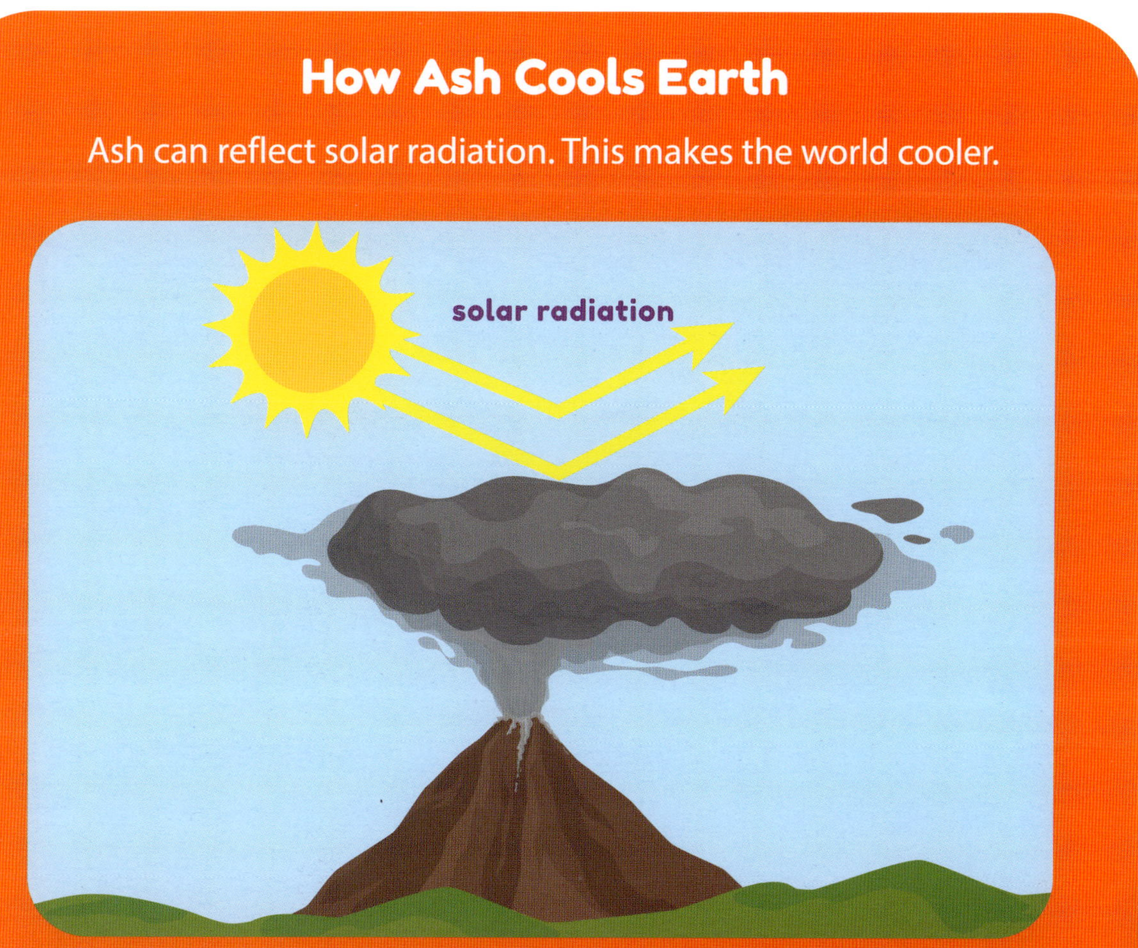

Droplets

Volcanic gases can also travel high into the atmosphere. The gases combine with water. They form tiny particles called droplets. These droplets reflect more radiation than ash does. This means they cool Earth more than ash.

FUN FACT!

Droplets can stay in the atmosphere for up to three years.

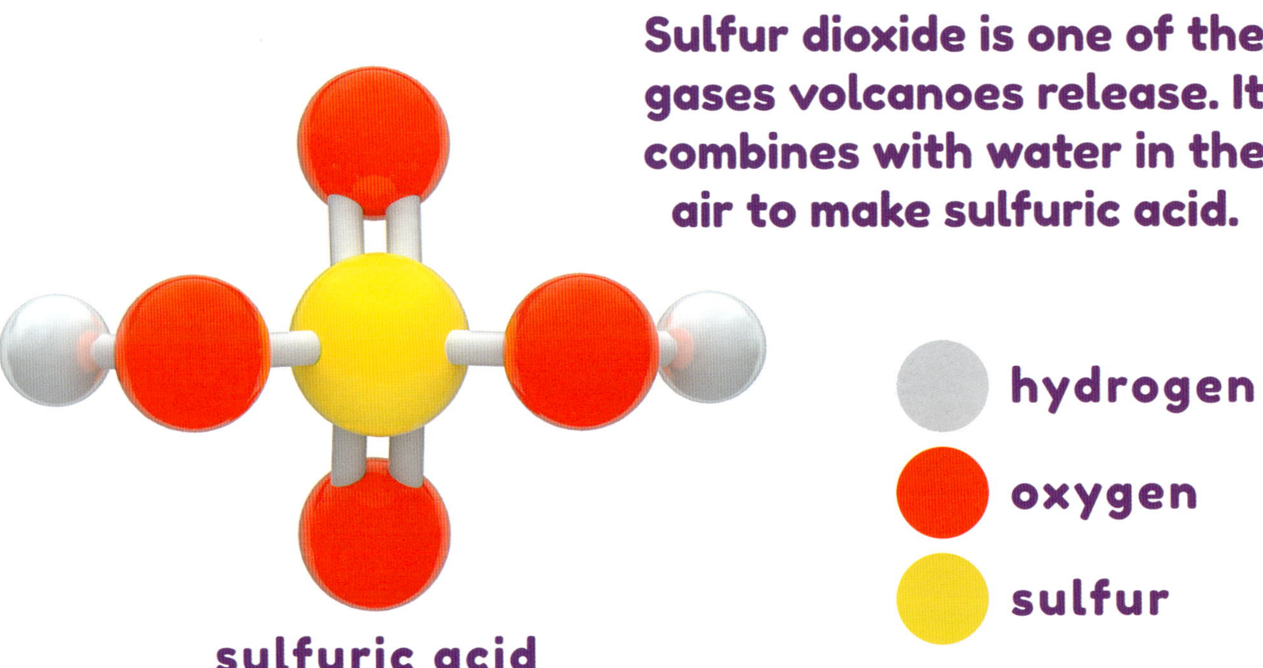

Sulfur dioxide is one of the gases volcanoes release. It combines with water in the air to make sulfuric acid.

Carbon dioxide is released naturally during forest fires and volcanic eruptions.

Carbon Dioxide

One gas volcanoes release is carbon dioxide. This gas acts like a blanket in the atmosphere. In large amounts, it traps Earth's heat. This can cause temperatures to rise.

Volcanic Winter

Droplets and ash can cause Earth to become a few degrees cooler. This period of cooling is known as a volcanic winter. Volcanic winters can last for months or years.

Burning fuels such as oil and coal releases carbon dioxide.

Volcanoes and Carbon Dioxide

Though volcanoes release some carbon dioxide, they do not make Earth significantly hotter. Human activities have a much bigger effect. Every year, humans release at least 60 times more carbon dioxide than Earth's volcanoes, causing temperatures to rise. This is called global warming.

More Eruptions

Volcanoes may not affect global warming, but global warming can affect volcanoes. Global warming causes glaciers to melt. This changes Earth's surface pressure. Surface pressure changes lead to more volcanic eruptions.

As global warming worsens, scientists study how the changing climate will affect volcanic activity.

A Quiet Volcano

Mount Vesuvius is a volcano in southern Italy. It is a composite volcano. Nearly 2,000 years ago, the volcano towered over the cities of Pompeii and Herculaneum.

Italy is a country in Europe.

Between 10,000 and 20,000 people lived in Pompeii before Mount Vesuvius erupted.

The Eruption

Pompeii and Herculaneum were ancient Roman cities. Pompeii was five miles (8 km) from Mount Vesuvius. Herculaneum was three miles (5 km) away. On August 24 in the year 79 CE, magma burst through the volcano's vent. A few hours later, the volcano erupted explosively. The entire top of the volcano blew off.

Many people in Pompeii were killed by falling debris.

Dangerous Cloud

The eruption created a large cloud of ash and rock. The cloud rose 27 miles (43 km) into the sky. Then it spread out in the atmosphere in an umbrella shape. It blocked out the sun. The tephra in the cloud rained down on Pompeii for 12 hours. Many pieces were up to three inches (8 cm) around. Some people fled. Others hid in their homes.

Herculaneum Destroyed

That night, the huge cloud became too heavy. It collapsed. It sent a wave of hot ash, gas, and tephra toward Herculaneum. Everyone in the city died. Volcanic mud and rock covered the city. It was more than 60 feet (18 m) deep.

The deep volcanic mud kept Herculaneum well preserved.

Pyroclastic Surges

Pyroclastic surges sped down the sides of the volcano. Poisonous gas flooded Pompeii. Everyone still in the city died. The last two pyroclastic waves were the most powerful. They brought massive amounts of rock and ash. Buildings collapsed. Pompeii was buried under 14 to 17 feet (4 to 5 m) of ash and rock.

FUN FACT!

About one-third of Pompeii is still buried in ash.

Some people in Pompeii died in ash. Their bodies decayed, leaving holes in the hardened ash. Scientists poured plaster in the holes. The plaster formed casts of the people who died.

Scientists are still making discoveries in Pompeii.

The Buried City

In 1748, a farmer found parts of Pompeii. Archaeologists began digging it up. The ash and tephra preserved Pompeii. The remains of the city show historians what life was like in ancient Rome.

Plinian Eruptions

A teenager named Pliny survived the 79 CE eruption. He wrote about the eruption of Mount Vesuvius. Today, explosive eruptions that create massive columns of ash are called Plinian eruptions.

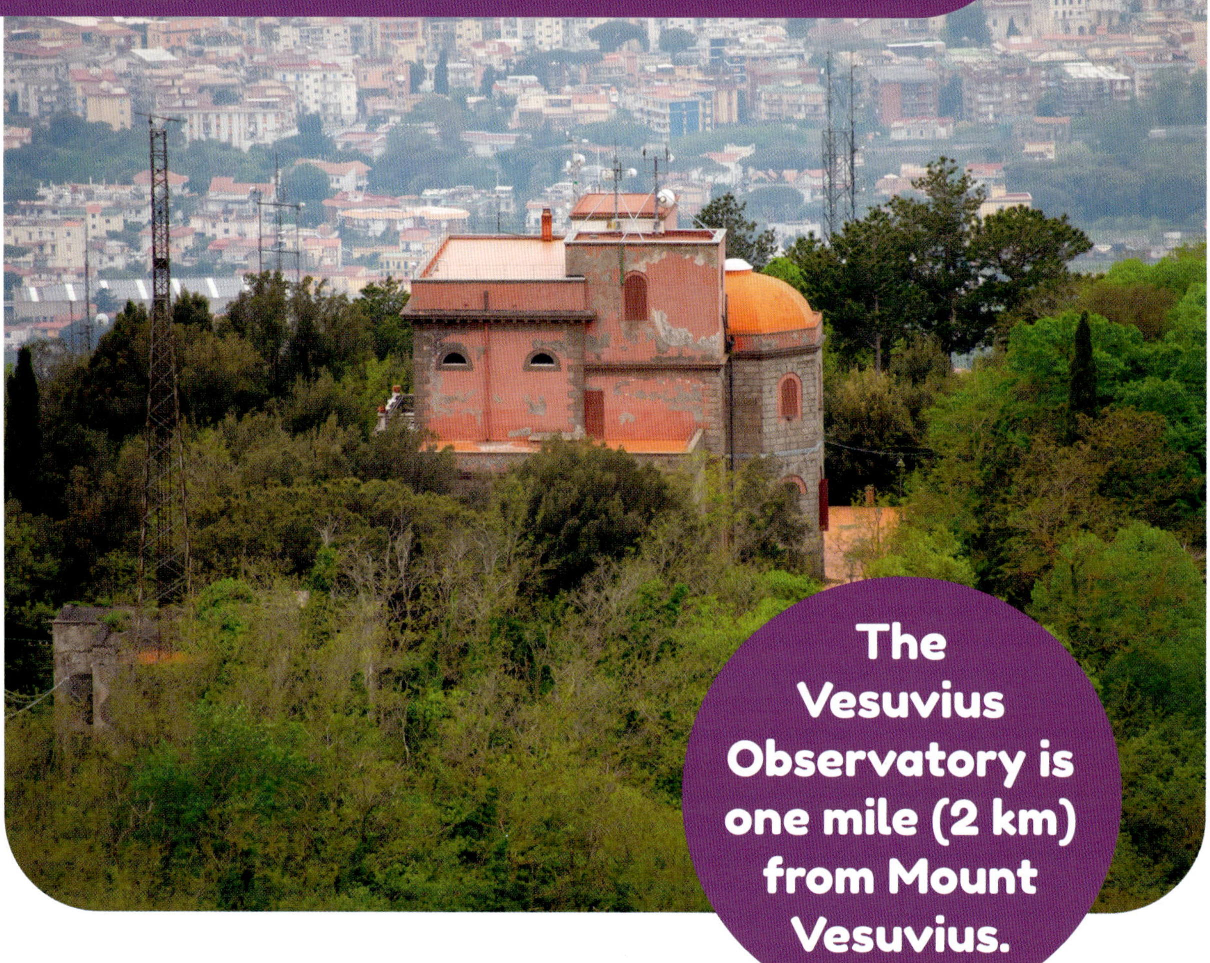

The Vesuvius Observatory is one mile (2 km) from Mount Vesuvius.

Studying Vesuvius

In the 1800s, the world's first volcano observatory was built near Mount Vesuvius. Scientists still study the active volcano today. The volcano has had 27 significant eruptions since it buried Pompeii. Its last major eruption was in 1944.

Dangerous Ground

Today, many people still live near Mount Vesuvius. The nearby city of Naples has a population of three million people. About 800,000 people live on the volcano's slopes.

Millions of people travel to Italy every year to visit the lost city of Pompeii.

Mount Tambora

Mount Tambora is a volcano in Indonesia. It is located on Sumbawa Island. It used to be 14,100 feet (4,300 m) high. It was one of the tallest mountains in Indonesia. On April 5, 1815, it began to erupt. On April 10, Tambora exploded.

Indonesia is a country in Southeast Asia.

Before Tambora erupted, the mountain was shaped like a cone.

Tambora Erupts

Fire, smoke, and gas blasted into the atmosphere. They went up 25 miles (40 km). The fire produced strong winds. The winds knocked down trees. Ash poured down the side of the mountain. It burned grasslands and forests.

A Distant Sound

The island of Java is hundreds of miles away from Mount Tambora. Soldiers in Java heard the eruption. They thought it was cannon fire. They went searching for a battle.

THE ERUPTION OF MOUNT TAMBORA

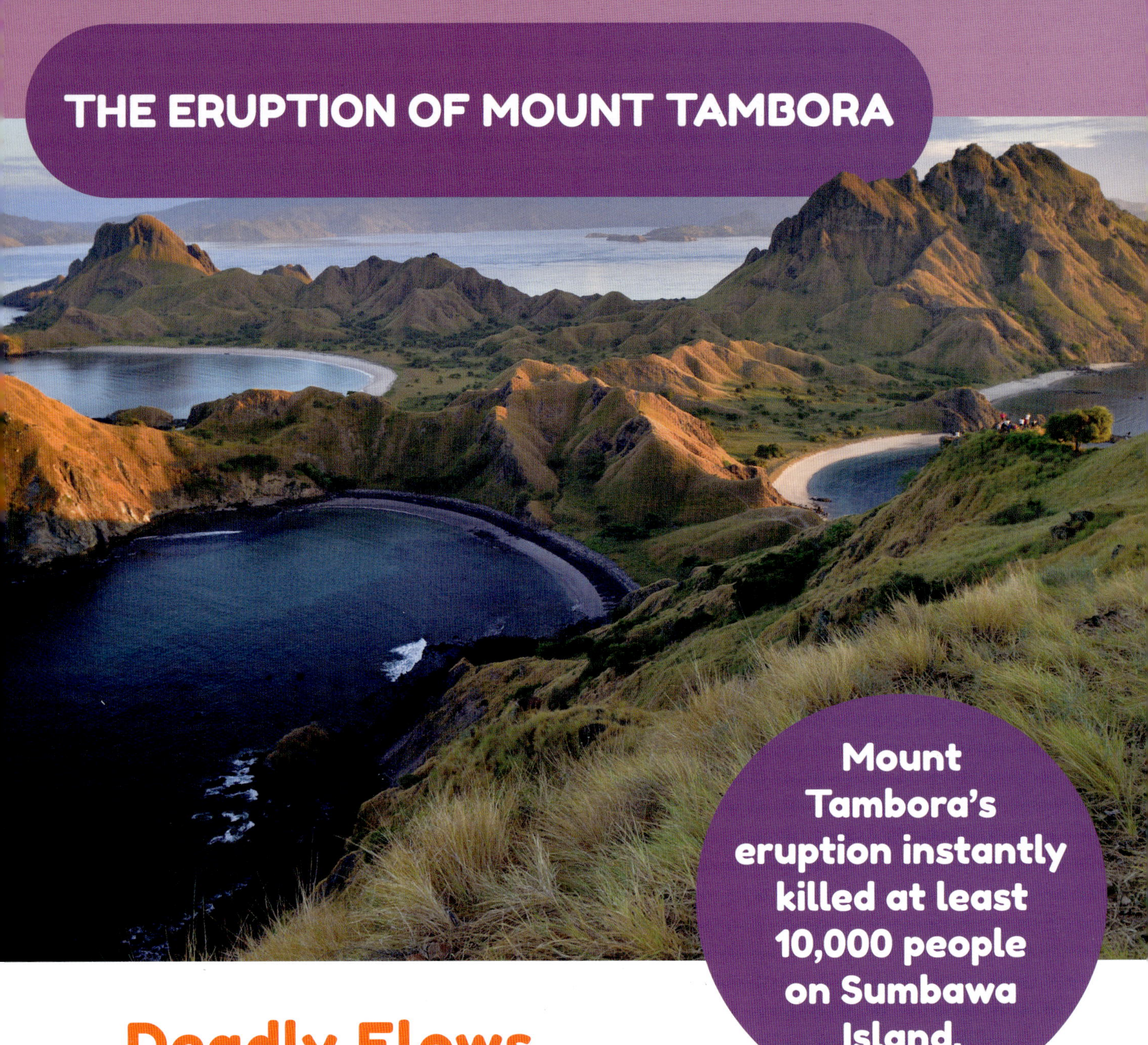

Mount Tambora's eruption instantly killed at least 10,000 people on Sumbawa Island.

Deadly Flows

Pyroclastic flows raced down the volcano's slopes. They traveled at more than 100 miles per hour (160 kmh). They destroyed everything in their path.

The Largest Eruption

Tambora's eruption was the largest volcanic eruption ever recorded. It was also one of the most destructive. Houses collapsed under ashfall. Fresh water became contaminated. Poisonous gases filled the air. Earthquakes shook the ground. They created tsunamis in the Java Sea.

Ash rained down on the areas around Mount Tambora for weeks after the volcano's eruption.

Long-Term Effects

Tambora's eruption ended in July 1815. But the effects continued for years. Small ash particles filled the atmosphere. They blew around the planet. On average, the ash made the global temperature drop by about 5.4 degrees Fahrenheit (3°C).

Mount Tambora Ashfall

Mount Tambora's ash devastated nearby islands and eventually spread around the world.

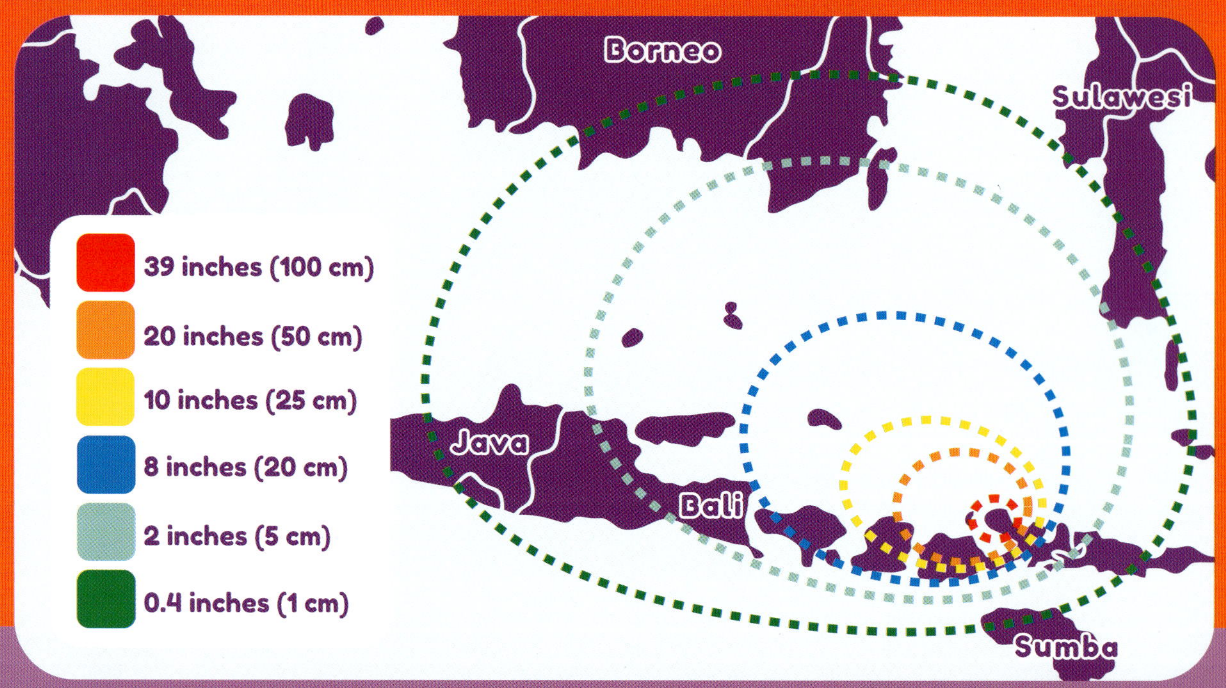

Despite Tambora being an active volcano, many people still live around the mountain.

Volcanic Winter

The cooler temperatures lasted for three years. They affected countries all over the world. Weather systems changed completely. The year 1816 became known as the Year without a Summer.

FUN FACT!

In the northeastern United States, weather changes from Tambora's eruption caused snow to fall in July.

The villages destroyed by Mount Tambora's eruption are buried under ash and pyroclastic flow. Some scientists are working to uncover them.

Starvation

The cooling temperatures decreased the amount of rainfall. This caused crops to die. People around the world began to starve. There was not enough food. The food shortage lasted for years. More than 100,000 people died as a result. This makes Tambora's eruption the deadliest eruption in history.

The Volcano Today

The eruption took off Tambora's top. Tambora lost more than 4,000 feet (1,200 m). Today it is just under 10,000 feet (3,000 m) tall.

Slow News

In 1815, there were no telephones. The telegraph had not been invented yet. News about the eruption traveled slowly.

Today, people can climb Mount Tambora and view the extraordinary volcano.

A Snowcapped Mountain

Mount Saint Helens is in the US state of Washington. It is part of the Cascade Range. It is a composite volcano.

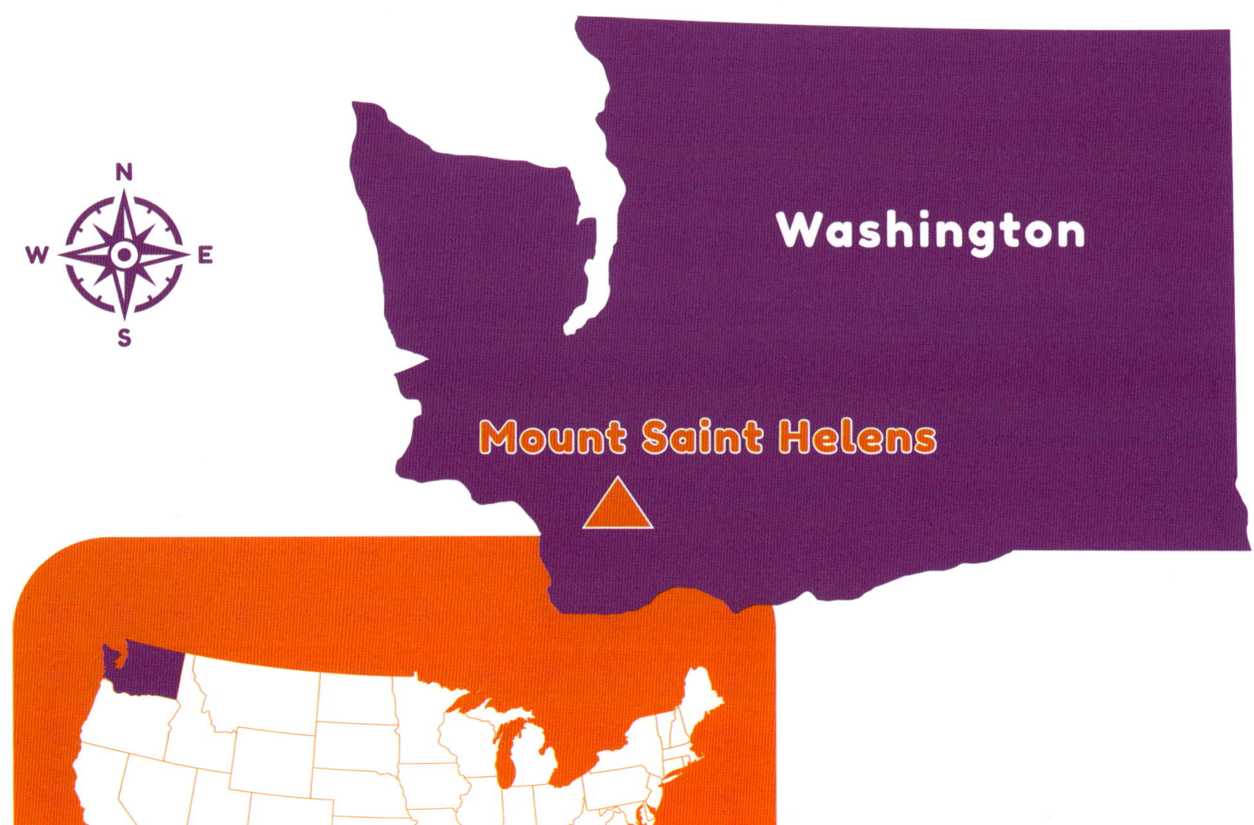

Washington is in the northwestern United States.

Mount Saint Helens's eruption on March 27, 1980, was its first eruption in more than 100 years.

A Crater Forms

On March 20, 1980, small earthquakes shook the north side of the mountain. On March 27, Mount Saint Helens erupted. Steam and ash exploded from the vent. The eruption created a huge crater.

Geologists traveled to Mount Saint Helens to investigate the first eruption. One was killed during the second eruption.

Earthquakes

In May, thousands of small earthquakes happened. There were hundreds of small explosions. Steam blasted from the volcano's vent. Meanwhile, the ground swelled. It created a bulge. The bulge grew to 450 feet (137 m). Scientists knew magma was rising.

Landslide

On May 18, a powerful earthquake shook the volcano. The north side of the volcano collapsed. It started a massive landslide. The mass slammed into a lake. It crossed a ridge. Then it rushed 14 miles (23 km) down a river. It was the largest landslide in recorded history.

The eruption did more than $1.1 billion in damage.

A Volcanic Blast

The landslide released gases in the volcano. The gases exploded. This created a deadly blast of hot gas, steam, and rock. The blast moved at about 680 miles per hour (1,100 kmh). The temperature reached around 570 degrees Fahrenheit (300°C). Pyroclastic flows followed, reaching 800 degrees Fahrenheit (430°C).

Mount Saint Helens's ash plume rose 80,000 feet (24,000 m) into the air.

Volcanic blasts that explode sideways are called lateral blasts.

Total Destruction

The volcanic blast tore through nearby forests. Approximately ten million trees were knocked down. It destroyed nearly all plant life within 12 miles (19 km).

FUN FACT!

The area of forests destroyed by Mount Saint Helens is known as the blowdown zone

Within 15 days, ash from Mount Saint Helens's eruption had spread around the entire Earth.

Ash Rises

At the same time, an umbrella-shaped column of ash rose from the vent and covered the sky. It turned everything dark. The wind carried ash over an area of 22,000 square miles (57,000 sq km). By the time the volcano calmed, 57 people had died. Most had been suffocated by ash or badly burned.

Still a Threat

Mount Saint Helens is about 1,300 feet (400 m) shorter than it was before its 1980 eruption. Since then, there have been more small earthquakes and eruptions. It is still considered one of the most dangerous volcanoes in the United States.

Today, Mount Saint Helens is protected as a National Volcanic Monument.

GLOSSARY

archaeologist
A scientist who studies human history.

atmosphere
The layer of gases surrounding Earth.

contaminated
Exposed to something harmful.

debris
The remains of something broken or destroyed.

downwind
In the direction that the wind is blowing.

evacuate
To remove someone from a dangerous place to a safer place.

fertile
Being able to produce lots of vegetation or crops.

glacier
A large, slow-moving mass of ice.

livestock
Animals kept or raised on a farm, such as cattle, sheep, or pigs.

molten
Something that has been turned into a liquid state due to being heated.

observatory
A place where scientists can monitor natural events or objects, such as volcanoes.

particle
A tiny piece of material.

radar
A system that uses radio waves to detect objects.

respiratory
Related to the lungs and breathing.

satellite
A human-made device that orbits Earth or another body in space.

3D
Three-dimensional, showing height, width, and depth.

TO LEARN MORE

More Books to Read

Gibbons, Gail. *Volcanoes*. Holiday House, 2023.

Murray, Julie. *Volcano Geology*. Abdo, 2023.

Van Rose, Susanna. *Volcano & Earthquake*. DK, 2022.

Online Resources

To learn more about volcanoes, please visit **abdobooklinks.com** or scan this QR code. These links are routinely monitored and updated to provide the most current information available.

INDEX

PHOTO CREDITS

Cover Photos: Ralf Lehmann/Shutterstock Images, front (lava); Photo by Gianni Sarasso/Moment/Getty Images, front (volcano); Evgeny Haritonov/Shutterstock Images, front (volcanologist); David Hogan/Moment/Getty Images, back

Interior Photos: iStockphoto, 1, 20, 97 (volcano), 104; Shutterstock Images, 3, 6, 8, 9, 11, 12, 13 (top), 13 (bottom), 17, 18, 24 (bottom), 26, 33, 34, 35, 38 (top), 45, 47 (top), 48, 53, 66, 67 (top), 67 (bottom), 69, 70, 72, 73 (top), 73 (bottom right), 87, 88 (top), 88 (bottom), 90 (top), 96, 97 (sun), 98, 99 (bottom), 103, 107 (bottom), 108, 111 (top), 111 (bottom), 117 (top); Henryk Welle/Moment/Getty Images, 4; Jim Sugar/Corbis Documentary/Getty Images, 5; Moisés Castillo/AP Images, 7; Red Line Editorial, 10, 29, 82, 83, 102, 110, 114, 118; Kriswanto Ginting/Moment/Getty Images, 14; Ralf Lehmann/Alamy, 15; Arctic-Images/Stone/Getty Images, 16; Gary Cook/Alamy, 19; Salvatore Allegra Photography/Moment/Getty Images, 21; Bayu Novanta/Xinhua News Agency/Getty Images, 22; Thorir Ingvarsson/iStockphoto, 23; Robert Schneider/Alamy, 24 (top); M. Hanum/Shutterstock Images, 25; Richard Roscoe/Stocktrek Images/Getty Images, 27; Photo by Gianni Sarasso/Moment/Getty Images, 28; Marli Miller/UCG/Universal Images Group/Getty Images, 30; Wolfgang Kaehler/LightRocket/Getty Images, 31; Deni Sugandi/Shutterstock Images, 32; Aditya Saputra/SOPA Images/LightRocket/Getty Images, 36; Martin Bernetti/AFP/Getty Images, 37; Fabrizio Villa/Getty Images News/Getty Images, 38 (bottom), 40, 51; Uldis Knakis/500px/Getty Images, 39; NASA, 41, 43; Firdia Lisnawati/AP Images, 42; Salvatore Laporta/KONTROLAB/LightRocket/Getty Images, 44; Martin Puddy/Stone/Getty Images, 46; Adhitya Hendra/Pacific Press/Light Rocket/Getty Images, 47 (bottom); Jacob Lowenstern/USGS, 49; USGS, 50, 77, 79; Andrew Richard Hara/Ena Media Hawaii/Getty Images News/Getty Images, 52; Charism Sayat/AFP/Getty Images, 54, 95; Luis Soto/AP Images, 55; moodboard/Image Source/Getty Images, 56; Kike Rincon/Europa Press News/Getty Images, 57; D. Damby/USGS, 58; Ignacio Palacios/Stone/Getty Images, 59; Ronny Adolof Buol/AFP/Getty Images, 60; US Geological Survey/Getty Images News/Getty Images, 61; Luis Robayo/AFP/Getty Images, 62; Guryanov Andrey/Shutterstock Images, 63; Europa Press News/Getty Images, 64; Dana Stephenson/Getty Images News/Getty Images, 65; Jeremy Woodhouse/Photodisc/Getty Images, 68; constantgardener/E+/Getty Images, 71; Olexandr Panchenko/Shutterstock Images, 73 (bottom left); Tanya Sid/Shutterstock Images, 74; David Hogan/Moment/Getty Images, 75; Paul Souders/DigitalVision/Getty Images, 76; Sholaita Iriawan/Pacific Press/LightRocket/Getty Images, 78; A. Darold/USGS, 80; Bernard Meric/AFP/Getty Images, 81; Puripat Lertpunyaroj/Shutterstock Images, 84; Ulet Ifansasti/Getty Images News/Getty Images, 85; Rakel Osk Sigurda/NordicPhotos/Getty Images News/Getty Images, 86; Orlando Estrada/AFP/Getty Images, 89; Giles Clarke/Getty Images News/Getty Images, 90 (bottom); James Davis Photography/Shutterstock Images, 91; Juan Mabromata/AFP/Getty Images, 92; Kevin Beavis/Alamy, 93; Jude Limage/500px/Getty Images, 94; Athit Perawongmetha/Moment/Getty Images, 99 (top); Kevin Frayer/Getty Images News/Getty Images, 100; L. DeSmither/USGS, 101; Kelly Cheng/Moment/Getty Images, 105; George Pachantouris/Moment/Getty Images, 106; Silvia Bazzicalupo/Anadolu Agency/Getty Images, 107 (top); Kenneth L. Kelly, 109; TeeJe/Moment/Getty Images, 112; Sueddeutsche Zeitung Photo/Alamy, 113; Khafid Mukriyanto/Alamy, 115; Fikria Hidayat/KOMPAS Images/AP Images, 116; M. Rinandar Tasya/Shutterstock Images, 117 (bottom); Sunset Avenue Productions/DigitalVision/Getty Images, 119; R. P. Hoblitt/USGS, 120; John T. Barr/Hulton Archive/Getty Images, 121, 124; HUM Images/Universal Images Group/Getty Images, 122; Bettmann/Getty Images, 123; George Ostertag/Alamy, 125